CREATED IN JAPAN

CREATED IN JAPAN

From Imitators to World-Class Innovators

SHERIDAN M. TATSUNO

Harper & Row, Publishers, New York
BALLINGER DIVISION
Grand Rapids, Philadelphia, St. Louis, San Francisco
London, Singapore, Sydney, Tokyo, Toronto

International Standard Book Number: 0-88730-373-0
Library of Congress Catalog Card Number: 89-38066
Printed in the United States of America

Library of Congress Cataloging-in-Publication Data

Tatsuno, Sheridan M.
Created in Japan: from imitators to world-class innovators/Sheridan M. Tatsuno.
p. cm.
Includes bibliographical references.
ISBN 0-88730-373-0
1. High-technology industries—Japan. 2. Creative ability in business—Japan. 3. Research, Industrial—Japan. I. Title.
HC465.H53T375 1989
338.0952—dc20 89-38066
CIP

89 90 91 92 HC 9 8 7 6 5 4 3 2 1

To Professor Shoei Ando,

my dear friend and teacher,

for opening my mind to Zen.

CONTENTS

LIST OF FIGURES *ix*

LIST OF TABLES *xi*

FOREWORD *xiii*

ACKNOWLEDGMENTS *xv*

PART I

Creativity Reconsidered

1 *The Sound of One Hand Clapping* 3

2 *East Meets West: The Yin and Yang of Creativity* 14

3 *The Creative Samurai: Japanese Intrapreneurs and Small Businesses* 26

PART II

Japanese Approaches to Creativity

4 *Historical Origins of Japanese Creativity* 41

5 *Sairiyo: Recycling the Past* 61

6 *Tansaku: Exploring New Ideas* 73

7 *Ikusei: Nurturing Creative Ideas* 87

8 *Hassoo: Generating Breakthroughs* 103

9 *Kaizen: Refining Ideas* 117

PART III

In Search of Creative Breakthroughs

10 *High-Definition Television: The Next-Generation Video Battlefield* 129

11 *The Computer Bazaar of the Future* 149

12 *Visions of Superconductors* 194

13 *Satori in the Laboratory: The Challenges Facing Japanese Researchers* 218

14 *Whither Japan in the Twenty-first Century?* 261

BIBLIOGRAPHY 275

INDEX 285

ABOUT THE AUTHOR 297

LIST OF FIGURES

2–1	U.S. vs. Japanese Modes of Creativity	19
2–2	The Japanese Approach to Creative Research	20
3–1	Japan's Corporate Restructuring	30
4–1	The Mandala of Creativity	51
4–2	The Evolution of Creative Japanese Ideas	54
5–1	The Japanese Approach to Hybrid Technologies	64
5–2	Interindustry Technology Fusion Index	65
5–3	Sanyo's Technology Fusion Approach to Bioelectronics	67
5–4	Television Spin-off Markets	70
6–1	Comparison of U.S.–Japanese Corporate Technology Search Approaches	77
6–2	Sharp's Digital Audio Technology Tree	83
6–3	Sharp's Audio Technology Road Map	84
7–1	Japanese Technopolis Areas	96
7–2	The Regional Research Core Program	99
8–1	The Lotus Blossom Method Diagram	112
10–1	The HDTV Mandala	148
11–1	The TRON Project	157

11–2 A TRON Keyboard Unit 160

11–3 The Supercomputer Project, 1981–89 163

11–4 Structure of a Neurocomputer 186

13–1 Key Technology Center Funding 241

LIST OF TABLES

1–1 The Evolution of Japan's Electronics Industry 6

2–1 The Yin and Yang of Creativity 16

4–1 The Cultural Foundations of Japanese Creativity 57

7–1 Regional Research Core Projects 100

10–1 MPT's Hi-Vision Market Forecast 137

10–2 MITI's Hi-Vision Market Forecast 138

10–3 Public and Industrial Hi-Vision Applications 139

10–4 Major Japanese Hi-Vision Promotional Events 140

11–1 Commercial High-Performance Scientific Supercomputers 165

12–1 Breakthroughs in High-Temperature Superconductivity 195

12–2 Japanese Government Superconductivity Funding (fiscal 1988) 200

12–3 STA's Superconductor Multicore Project (fiscal 1988) 201

12–4 Research Divisions of the International Superconductivity Research Laboratory 203

12–5 Superconductor Market Forecast by Japanese Experts 209

13–1 Joint University–Industry Research Centers Sponsored by MOE 230

13–2 New Foreign University Campuses 231

13–3 JRDC's ERATO Program 237

13–4 Japan's Major Large-scale R&D Projects 239

13–5 Overseas Japanese R&D Centers 244

13–6 New Foreign R&D Centers in Japan, 1986–89 248

13–7 Japanese Investments and Acquisitions of High-Tech Companies 257

13–8 MIT Chairs Sponsored by Japanese Corporations 259

FOREWORD

SINCE the Meiji Restoration over 120 years ago, Japan has been trying to catch up to Western living standards by modernizing its industrial base. Japanese industry has had to meet the additional postwar challenges of contributing to the rehabilitation of a war-devastated nation and reducing Japan's large international debt. By emphasizing quality control and effective sales techniques as much as technology development, modern Japanese industry has mainly succeeded in these efforts.

Now motivated by two new factors, Japan is developing its own technical knowledge that can contribute to the global economy. The first is Japan's achievement of almost complete parity with the United States and Europe in industrial technologies. Copying Western technologies has become an outmoded way of conducting business; Japan must now develop innovative new technologies and markets itself in order to become a world leader. Second, industrial society is quickly being transformed into a global information society, a change that requires new, creative R&D efforts from all nations, including Japan.

The key to competing successfully in the global market is overall strength. Each link in the chain—from basic research to R&D to production to marketing—must be as strong as the next. Western Europe may do better basic research than Japan, and the United States may do better R&D. But I believe that overall strength is the secret to the current success of Japanese industry.

Sheridan Tatsuno has thoroughly studied the forms of Japanese creativity behind the increased international competitiveness of Japanese industry, especially the electronics industry. In our information-oriented world, major technological innovations are no longer based on single technologies; as Mr. Tatsuno points out, technology fusion will play an important role in new discoveries and breakthroughs in the 1990s. Since Japanese industry has already become a dominant force in technology fusion, Japanese creativity is a particularly worthy subject for study. The publication of this book could not be more timely.

—DR. MICHIYUKI UENOHARA
Executive Vice President and Director
NEC Corporation

ACKNOWLEDGMENTS

Books are passions of the mind,
conceived in the heat of debate.

IN the summer of 1987, on a return flight to Silicon Valley from a speaking engagement at a venture capital conference in Aspen, Colorado, I was reliving a conversation when these words popped into my head. That morning, I had exchanged a few parting thoughts with Dr. Terry Winters, general partner of Columbine Ventures, who was not persuaded by my argument that the Japanese were becoming more creative.

"If the Japanese are so creative," he maintained, "why haven't they made more scientific breakthroughs or won more Nobel Prizes? Even West Germany, which was also devastated in the war, has made more significant discoveries."

To drive home his point, he sent me a copy of a *Wall Street Journal* article by Stephen Kreider Yoder, who deplored the lack of scientific creativity in Japan. "I rest my case," Dr. Winters penned in the margins.

Not given to conceding easily, I took his words as an intellectual challenge. If the Japanese are not creative, how could one explain the freshness and imagination of their architecture, product design, and photography? Or the originality of the Sony Walkman? Although most Westerners believe the Japanese to be mere imitators, I have seen too much evidence to the contrary. Two books, *The Best of Japan* and *283 Useful Ideas from Japan,* confirmed my suspicions. Perhaps Western stereotypes about Japan needed to be challenged; perhaps there would be more than meets the eye if one looked below the surface. Deliberately taking a contrary viewpoint might reveal new insights.

My search for answers uncovered a wealth of ideas, mostly by Japanese commentators. For some reason, Western business books seem fixated on Japan's past, not its future. Except for one book, *Creativity in Business* by Michael Ray and Rochelle Myers, and a few scattered articles, there are few references in business journals or magazines to Japan's R&D spending boom, its shift to creative research, or Eastern forms of creativity. Fortunately, new journals such as *Look Japan* and *Business Tokyo* are filling the void.

Insights from several friends and colleagues indicated that I was on the right track. William Watson, president of National Japan and a longtime "Japan hand" in Tokyo, pointed out fields in which he found the Japanese very creative. And Keisuke Yawata, president of LSI Logic K.K., who has years of experience on both sides of the Pacific, was extremely helpful in describing the differences between Japanese- and American-style creativity. Their comments were reinforced by Katsuhide Hirai, director of information systems at Fujitsu America, who described for me the basic differences he finds between the Japanese and American approaches to software development at Fujitsu. Dr. Michiyuki Uenohara, a former Bell Labs researcher and director of R&D at NEC Corporation, described his search for creative new R&D management techniques.

My special gratitude goes to Frederik L. Schodt, author of *Inside the Robot Kingdom,* a fascinating look at Japan's robot industry. Mr. Schodt introduced me to Sueo Matsubara, acting director of the Mukta Research Institute in Tokyo, by whom my eyes were opened to an emerging world of Japanese management literature based on the precepts of Buddhism. In particular, I am indebted to Mr. Matsubara for broadening the philosophical perspective of my "mandala of creativity": the creative process can be conceived of as a helix through which products, like people, are recycled again and again, each time becoming a little bit better and purer—a form of *satori* through product development. Although we have not met, my thanks also go to Masahiro Mori, Tokyo University professor and founder of the Japan Robotics Association, whose theories on nonduality in thinking have guided the members of the Mukta Institute. Diane Yoshikawa, director of the Osaka–California Linkage (OCL) Center, provided invaluable comments on the role of nondual thinking in Japanese society.

Throughout the writing of this book, I have been deeply influenced by Yasuo Matsumura of Clover Management Research, whose MY Method—which I call the "lotus blossom" technique—is used by many Japanese businesses for product development. Recently, the Dentsu advertising agency distributed an "HDTV Mandala" poster based on the MY Method to educate the Japanese public about high-definition television.

For the last two years, this book drifted among the emerging management ideas in Japan, but remained unanchored in the West. For their help in steering my research home, I am deeply appreciative of Paul Saffo and Bob Johansen of the Institute for the Future who helped me see the conceptual link between my research findings and management studies in the West.

Finally, I would like to thank Marjorie Richman, my original editor at Ballinger, and Dataquest, for granting me permission to use its information for this book. Without their encouragement and support, this book would not have been possible.

Fremont, California
September 1989

PART I
Creativity Reconsidered

1

The Sound of One Hand Clapping

ZEN MASTER: What is the sound of one hand clapping?
PUPIL: Your own hand.

CONSIDER the following: In 1976 the top four recipients of U.S. patents were General Electric, the U.S. Navy, Bayer, and Xerox. By 1987 three of the top four were Japanese companies—Canon, Hitachi, and Toshiba. The U.S. Patent and Trademark Office reported that foreigners received half of the 89,385 U.S. patents issued in 1987. The Japanese share was 20 percent, and growing. Moreover, a 1988 National Science Foundation (NSF) study suggests that Japanese patents are more innovative than U.S. patents: researchers found that, since 1976, U.S. patents held by Japanese have been cited by other researchers more often than patents awarded to Americans.

At the International Solid State Circuits Conference—the Olympics of semiconductor research—the United States accounted for 61 percent of the leading technical papers in 1981. The Japanese share was 25 percent. By 1987 the Japanese had garnered 44 percent, while the U.S. share had dropped to 43 percent. That same year, the Defense Science Board concluded that Japan held a "clear and increasing lead in most silicon product technologies," its data showing Japan ahead in eleven equipment areas, the United States ahead in seven areas, and the Japanese gaining strength across the board.

In 1984 Pentagon officials visited leading Japanese research laboratories and petitioned the Japanese government for access to sixteen critical technologies that Japan had achieved parity with, or a lead over, the United States in developing.

In 1985 a General Accounting Office (GAO) report concluded that Japan had grabbed the lead in commercial optoelectronics and was gaining ground in optical communications.

In 1986 the Houston Research Council purchased NEC's SX-2 supercomputer over the leading U.S. model, citing better performance for the cost. The White House recently announced a national supercomputer strategy to maintain U.S. competitiveness vis-à-vis Japan.

The Silicon Valley market research firm Dataquest reported in 1989 that Japanese companies captured 50 percent of the $50 billion global semiconductor market in 1988, while the U.S. share dropped to 36 percent. NEC, Toshiba, and Hitachi were the leaders. Silicon chips—the "brains" for America's high-tech future—will be a $96 billion global market by 1996.

In 1988 the NSF reported that the Japanese had achieved parity in superconductivity research and were pulling ahead of the United States in commercial applications.

In 1989 the American Electronics Association warned that Japan was years ahead in high-definition television (HDTV)—the key to advanced computer graphics and chips in the 1990s. HDTV will become a pivotal market over the next twenty years, with sales exceeding $145 billion.

The Rise of Japanese Creativity

Japan's creativity in science and technology is on the rise. After spending decades mastering quality, Japan is now emphasizing basic research and creative "hit" products, and the effort is already paying off. Japanese patents are multiplying worldwide. New products such as Sony's Video Walkman, Minolta's Maxxum 7000, and Mazda's Miata have landed on our shores. Japanese researchers are making greater contributions at major scientific conferences. In industry after industry, from supercomputers to high-definition television and superconductivity, Japan is challenging or leapfrogging the West.

Signs of Japanese creativity are everywhere. At international trade shows, Japanese companies are trend-setters in industrial design, audiovideo equipment, computerized language translation, bullet trains, car navigation systems, advanced robots, and factory automation. New product ideas, such as electronic keyboards, ceramic paper, plastic dancing flowers, and floating factories, are emanating from Tokyo. Finding themselves no longer cost-competitive with the rest of Asia, Japanese companies are rushing into upscale markets for high value–added products and services, many of which have never been seen in the West. Walking into crowded bookstores in Tokyo can be an overwhelming experience. One is confronted by hundreds of books, both foreign and domestic, on new technologies, innovation, creativity, and research management.

Time magazine noted in March 1988: "The harsh truth is that if at one time the Japanese could be dismissed as mere imitators, that time is long gone." Bernard Wysocki, Jr., of the *Wall Street Journal* is equally frank: "All the while, though, we had one big consolation. The Japanese were essentially copycats. . . . No longer. Today Japan is challenging us even here, on the slippery rockface of new ideas, innovations, new technologies. Having caught up with the U.S. in many older technologies, Japan is struggling to pass the U.S. in new ones."

From Quality to Creativity

Japan's transformation from imitator to world-class innovator has been a dramatic story unparalleled in the postwar era (see Table 1–1). During the 1950s and 1960s, Japanese companies produced low-cost imitations; by the end of the 1960s, however, they had achieved the manufacturing quality and product reliability that are now acknowledged as global standards. "Quality" became the Japanese rallying cry, an emphasis that paid off handsomely in world markets.

During the 1970s, Japanese companies focused on product design and manufacturability. Instead of exporting jobs to Southeast Asia, as U.S. companies were doing, they maintained tight control over critical manufacturing processes and quality assurance. As the West's manufacturing capability weakened, Japanese companies invested heavily in state-of-the-art equipment and gained global market share in many industries. They made numerous product improvements, but were still criticized for reverse engineering and product imitation.

During the 1980s, however, innovation became Japan's new industrial slogan; imitation was no longer assuring corporate survival. Fierce global competition and the rising yen forced companies to find ingenious ways to recycle their technologies into new products, such as 8mm video cameras, personal facsimile machines, Japanese-language word processors, compact disk "jukeboxes," and laptop personal computers.

During the 1970s and 1980s the United States, on the other hand, squandered its technological lead. We became so engrossed with financial wizardry, short-term profits, fat military contracts, lucrative mergers and acquisitions, and paper entrepreneurialism that we consistently underinvested in long-term education, basic research, and training. As our economy became heavily militarized and inflated, our product quality plummeted, services deteriorated, and our educational stan-

TABLE 1–1
The Evolution of Japan's Electronics Industry

Period	Industry Emphasis	Management Technique	Target Industries
1950s–1960s	Quality	Total quality control (TQC)	TVs, radios, steel, chemicals
1970s	Aesthetic design	Product refinement	Cameras, audio equipment, VCRs
1980s	Product innovation	Spiral development	Laptop computers, 8mm video, facsimile machines
1990s	Creativity	Technology fusion	Bioceramics, neural networks, optomechatronics, bioelectronics, videocomputers
2000+	Spiritual and physical well-being	Humanware engineering	Biomechatronics, biocomputing, biocommunication

SOURCE: Dataquest, Inc.

dards hit rock bottom. Most important, we failed to notice our Pacific ally venturing out on its own. We preferred to believe the "feel-good" newspaper and magazine articles and management books telling us that Japanese companies are only good at mass production and that the United States was still ahead in basic research and product innovation. We fell asleep at the wheel: mesmerized by our own entrepreneurialism and creativity, we couldn't see the competition at our doorstep.

The Dangers of Japan-Denial

Is talk of Japanese creativity an exaggeration? Or has a decade of "standing tall" lulled us into dismissing Japan's creative effort as a feeble attempt to copy Yankee ingenuity? The Japanese are good at stealing and repackaging our ideas, goes this argument, but they aren't very creative, just clever copiers. After all, where are their Nobel laureates? Where is their creative software? Everyone knows that rugged individualism and the freedom to explore and make mistakes are necessary for creative genius to thrive. Japanese society is highly regimented and conformist; thus the Japanese cannot possibly be as creative as we are. Western industry may be under the gun, but Western creativity will always reign supreme over "Japan Inc."

In response to the Japanese challenge, many of our journalists and scholars fall back on these hackneyed stereotypes and myths about Japan without examining their validity. But does the West have a monopoly on creativity? While it is true that the Japanese have been scientific followers, they have demonstrated extraordinary creativity in many fields of endeavor: ceramics, video equipment, photography, fashion design, and architecture, not to mention their exploratory research in next-generation technologies, such as Kyocera's bioceramic bones and teeth, Fujitsu's neural computer, and Sony's high-definition TV research.

The comfortable stereotype of Japan-as-copycat no longer squares with the technological advances and risk-taking going on in Tokyo. Perhaps we are hearing the sound of one hand clapping—the sound of Americans patting themselves on the back. Or the sound of Japanese creativity falling on ears that are still deaf to new forms of creativity. Our cultural chauvinism—reflected in the Not Invented Here (NIH) syndrome ("If it wasn't invented here, it can't be very good")—is perhaps the main reason for our deafness. The West has been the center of scientific discoveries and technological breakthroughs for so long that most Westerners routinely dismiss signs of Japanese creativity, remaining too busy or too lazy to investigate further. In his book *Made in Japan,* Akio Morita, chairman of Sony, warns us about ignoring these signs:

> If you go through life convinced that your way is always best, all the new ideas in the world will pass you by. Americans tend to think that the American system is the way things should work all around the world, but they should not be blind and deaf to how things are done in other countries.

Since World War II, we have lost industry after industry to Japan because of our complacency, disinterest in foreign markets, and just plain arrogance. In the 1950s, we overlooked Japanese quality improvement efforts because, according to prevailing wisdom, the Japanese made shoddy products that were cheap imitations of ours. But Japanese companies were making millions of refinements and improvements that gradually led to major product innovations, such as the Sony Walkman. This is just one example of our blindness to Japanese creative efforts. Another is the fact that while leading-edge technologies in the United States are familiar to most Japanese researchers, technologies considered "old news" in Tokyo, such as superfast trains and biosensors, are viewed as "science fiction" by my colleagues in Silicon Valley, who seem totally unaware of Japanese developments in their own fields.

The language barrier has not helped matters much. Despite the creative outburst in Japan, very little in-depth reporting on it and few scientific papers are available in English or other

Western languages. Most coverage of Japan still tends to be superficial, rehashing the decades-old stereotypes. Because of the difficulty of mastering written Japanese—once called "the devil's tongue"—few Westerners have access to the phenomenal flow of scientific and technical information available in Japanese. As a result, most rely on watered-down summaries or self-serving translations from Japanese sources.

Our schools and companies have not trained our managers, engineers, and students to overcome these language and cultural barriers. Indeed, few of our Fortune 500 executives can speak foreign languages as fluently as their European, Latin American, Asian, or Japanese counterparts. As a result, they are often out of touch with the latest developments around the world. Unwilling to acknowledge that others might be ahead in key technologies, they close their eyes and ears to foreign announcements. Unfortunately, we are being beaten by Japan in global markets before we even begin to compete.

By contrast, all major Japanese corporations and newspapers closely follow technological, political, and economic developments around the world. In fact, it is often easier to monitor developments in the United States through Japanese newspapers and trade journals because of their exhaustive reporting. Japanese companies train their people in foreign languages and send teams of managers and researchers to see new technologies and products firsthand. The Japanese realize that creative ideas have no geographic boundaries.

Nor does our economic future, which increasingly is being created in Japan. Decisions about new plant investments, mergers, and acquisitions are made in corporate suites throughout Tokyo, financed by Japanese banks. New product strategies are first tested in Japan. Since the mid-1980s, Japanese companies have opened over 200 basic research laboratories in electronics and biochemistry, where researchers are exploring new technologies such as optocomputers, bioelectronics, neural networks, fuzzy computer logic, and translation phones. They are the "new Americans"—pioneering at the edge, fueled by their self-confidence and enormous wealth.

Forces Behind Japan's Transformation

Why are the Japanese pursuing creativity so feverishly? Since 1985, the global economy has changed dramatically, forcing Japanese companies to restructure and reposition themselves. The yen has doubled against the U.S. dollar and has risen over 50 percent against other major currencies, making many goods too costly to produce in Japan. To remain competitive, Japanese companies are laying off part-timers and subcontractors, spinning off new venture subsidiaries, retraining and reshuffling workers to high-growth sectors (such as compact disks, telecommunications, and software), and opening new plants and research centers overseas. Even the old boy network of suppliers and distributors is falling apart as Japanese companies import more and more parts from other Asian suppliers at lower cost.

As the yen strengthens, there is fear that Japan will rapidly be "hollowed out," or deindustrialized. Tough competition from South Korea, Taiwan, Hong Kong, and Singapore is forcing Japan to shift to new products. In the videocassette recorder (VCR) market Samsung and Daewoo are grabbing shares from Hitachi, Matsushita, and Sony at the low end, while in the auto market Hyundai's Excel has forced Honda, Nissan, and Toyota into upscale markets. And the pace of competition is accelerating. Already, Thailand, India, and China are "bumping" Taiwan and South Korea markets upstream into Japanese markets.

Tough enforcement of intellectual copyrights and patents is creating yet another pressure on Japanese companies to develop their own creative technologies. During the postwar period, Japan could benefit from the technological free ride provided by U.S. companies anxious to license their technologies for revenues. Now many Western companies refuse to allow Japanese researchers into their laboratories for fear of reverse engineering and the "boomerang effect"—that is, seeing their markets inundated with look-alike Japanese products. Recent concern over the leakage of advanced technologies to the Communist bloc through Japan's "leaky bucket" is preventing many

Western companies from sharing their technologies with Japan. Toshiba's sale of submarine propeller grinding equipment to the Soviets is a case in point.

Japanese corporations are realizing that they cannot survive without investing more in their own research and development (R&D). In the first half of the 1980s, corporate research staffs increased by 42 percent. In 1985 Japanese R&D spending reached 2.77 percent of the gross national product (GNP), exceeding that of the United States. The government's goal is to reach 3 percent by the early 1990s.

This heavy R&D investment has already paid off handsomely in a flurry of innovative new products. Hitachi launched the "Quiet Lady," a fully automatic washing machine for apartment dwellers who are concerned about making too much noise. Yamaha's digital WX7 looks like a clarinet but plays like a saxophone and can be made to sound like a trumpet, saxophone, piccolo, guitar, or violin. Cosmetic maker Kao Corporation used biotechnology to create "Attack," a laundry detergent as powerful as one-fourth the amount of conventional detergents; Kao racked up sales of $77 million in the first six months of sales to capture 30 percent of the Japanese market. Matsushita Electric and Funai Electric introduced an automatic bread baker in 1987 that was so popular that they immediately sold out; a $300 million market was created overnight. Personal facsimile machines—originally designed to accommodate the complex *kanji* writing system to the instant communication needs of the modern world—have transformed the way we conduct business. Sanyo Electric has developed the world's first see-through batteries—translucent amorphous solar cells for use as windows in greenhouses and skylights to absorb natural light and generate electrical power. Perhaps most novel of all is the "Toto Queen," an automated toilet with chips that direct the seat to preheat and to trigger a spray of bottom-washing water during flushing, followed by a blast of hot air and a puff of perfumed mist!

Japan is also faced with many internal pressures for change. Japanese society is rapidly aging, and its women, like American women before them, are entering the professional work force. The independent-minded young people, nicknamed the "new breed" (*shinjinrui*), are raising concern among senior managers about Japan's future competitiveness. These younger Japanese are tired of their workaholic existence and do not enjoy being "bashed" by foreign countries who talk about Japan's "economic miracle." Although Japanese incomes are now on average higher than U.S. salaries, few young Japanese can afford a decent home. The Japanese are now aspiring to the high living standards of the West. They want to enjoy life, to have more time for sports, travel, hobbies, arts, culture, and leisure activities. However, their desire for better lifestyles is being stifled by the Japanese government and businesses, which prevent them from buying lower-cost foreign imports due to cumbersome distribution systems and restrictions on foreign-owned wholesale outlets.

How will the Japanese respond to these new challenges? How will they meet changing global realities, as well as social changes transforming their country? What are the dilemmas facing Japan as its people become more creative, international-minded, and diverse? How will Japan's technological strength and creative blossoming affect its relations with other nations? Is Japan sitting on a gold mine or a political time bomb?

If the recent past is prologue, the Japanese will meet the challenge in a gradual, controlled manner. They will continue to create new technologies and will surpass us in dozens of industries during the 1990s—and the political reaction in the West will be deafening. But with its newfound wealth and creativity will come a host of new responsibilities. Will Japan be able to handle the pressures of being a technological leader?

We are witnessing a historic transformation. Japan is turning away from the West and reexamining its own cultural roots. It no longer seeks to copy the West, but to develop its own forms of creativity. The Heisei period may well become Japan's "second Renaissance," recalling the Heian period (710–794 A.D.) when Japan emerged from the shadows of Chinese culture and created its own culture.

Japan's push for renewed creativity will shake up relations between the world's interdependent nations. The United States may have been the scientific pioneer, but the baton is being passed to Japan, which is discovering new ways to meld Eastern and Western thinking into a new synergistic approach. Unless we wake up soon to the implications of Japan's latest creative endeavors, the West will be left in the dust as Japan blazes a trail of its own into the twenty-first century.

2

East Meets West: The Yin and Yang of Creativity

> Most Japanese creativity in electronics has been in the form of concepts for which the stimulus was already provided; they have been forcibly induced through concentrated effort. I will call this adaptive creativity.
>
> —Makoto Kikuchi
> *Japanese Electronics*

As we enter the 1990s, renewed Japanese creativity poses a series of seemingly unsolvable paradoxes, Zen puzzles for modern industry. What is Japanese creativity? How does it differ from Western creativity? How can it be fostered in Japan's conservative, risk-averse society? And how can it be used to revive dying industries and bolster high-tech businesses? Today these questions are being seriously discussed on both sides of the Pacific Ocean—not only because of Japan's technological strength, but because of its growing responsibility for helping to solve global problems such as toxic wastes and the greenhouse effect. The Japanese are returning to their cultural roots; they are seeking creative answers to new problems. When they unleash this latent creativity, we are likely to witness an explosion of ideas on a scale comparable to the Industrial Revolution. Indeed, the 1990s could well mark the beginning of a "creativity revolution." What will trigger this explosion in creative thinking?

Globalization is the driving force. Since World War II, air

individualistic, while the Japanese are much more adaptive, holistic, and cyclical in their thinking. Let's examine these differences one by one.

Breakthrough Creativity vs. Adaptive Creativity

Leading Japanese and American experts have different theories about Japanese creativity, but both sides agree on one point: Americans tend to excel at breakthrough research, while Japanese are better at adapting ideas and technologies to create new products and markets. Western creativity reflects our frontier thinking, while Japanese creativity is based on natural and social constraints.

According to Makoto Kikuchi, Director of the Sony Research Center, Westerners are stronger at independent creativity, which, being based on concepts of individual freedom, favors the discovery of new ideas and product breakthroughs. He feels that Japanese, by contrast, are better at adaptive creativity—refining ideas within the framework of new theories. This systematic approach is more suitable for process innovations and complex systems research, such as factory automation software or high-definition television, both of which require continuity and teamwork.

Professor Henry E. Riggs of Stanford University's Department of Industrial Engineering echoes this view of Japanese creativity. "Japan has demonstrated great abilities in the area of process technology, while the United States has been relatively more capable of product development." Riggs notes that, in almost all fields, the Japanese excel when research targets

TABLE 2–1
The Yin and Yang of Creativity

The Western Approach	The Japanese Approach
Breakthrough creativity	Adaptive creativity
Spontaneous creativity	Cultivated creativity
Creative fission	Creative fusion
Cartesian logic	Fuzzy logic
Uni-functional creativity	Multi-functional creativity

travel and communications have dramatically reduced the tance between nations. Instead of taking a two-week boat one can fly from San Francisco to Tokyo in ten hours. In national travel, once the domain of the wealthy, has bec during the 1980s a form of commuting. Telephones and facs ile machines are lessening Japan's geographic isolati Whether on business or vacation, the Japanese are exchang ideas with others as never before. There is a boom in overs offices, international joint ventures, and foreign companies ting up operations in Tokyo. The Japanese media reflects t growing curiosity in foreign countries. A popular Japanese evision program, "*Naruhodo* (I see!)—The World," explores t globe for fascinating ideas. This new mingling of cultures a ideas contains the seeds of creative change.

In the nineteenth century, the French mathematician Jul Henri Poincaré observed that a creative idea often emerg from the convergence of two totally unrelated ideas. For exa ple, the automobile resulted from placing a steam engine on a horseless carriage. The Wright brothers invented the airplan by putting an engine onto a glider. Mitch Kapor incorporate Eastern philosophy into his best-selling Lotus 1-2-3 software. B combining two different ideas, new industries were born. On broader scale, this creative fusion—the meeting of East an West—is the key to Japan's economic future. It is the yin an yang of creativity.

Unfortunately, many Westerners overlook Japanese crea tivity, believing Japanese society to be unconducive to Western style individual creativity. Those maintaining this ethnocentric view ignore the rich potential of group creativity, which Asian and other non-Western peoples have evolved over the centuries. Such an attitude forecloses the possibility of combining Western creativity with Japanese creativity to develop new ideas.

Differences in Western and Japanese Approaches to Creativity

The differences in Western and Japanese creativity are outlined in Table 2–1. Westerners tend to be linear, rational, and are clearly defined and fixed, whereas Americans do better when targets are vague, allowing for more personal freedom and individual creativity. Because they are driven by users and in-house suggestions, Japanese companies analyze the limitations of existing processes and make small improvements. They are incrementalists.

Adaptive creativity is a way of creatively responding to breakthroughs. For example, Japanese companies often recycle old customs and ideas, such as miniaturization or multiple use, and combine them with other ideas to develop new ones. For example, George C. Devol, Jr., and Joseph Engelberger pioneered the first industrial robots in the United States, but the Japanese are using the technology to develop piano-playing robots, sushi-making robots, fire-fighting robots, and hospital-care robots. Their emphasis is not on making breakthroughs but on humanizing and transforming these technologies for applications in everyday life. Japanese creativity is thus highly responsive, eclectic, focused, and practical.

Spontaneous Creativity vs. Cultivated Creativity

Why do the Japanese take an adaptive approach to creativity? Why are focus and control so important to them? To understand the cultural differences between Japanese and Western creativity, let's look at the analogy of farming in the two cultures.

Western creativity is like growing wheat, which cannot be forced but must be allowed to develop spontaneously in its own way after the proper conditions have been provided. The land must be plowed, the soil fertilized, and the seeds planted. But thereafter the process simply takes time; it cannot be speeded up by overfertilizing or by inspecting the plants each week. Similarly, Western ideas are allowed to germinate quietly on their own.

By contrast, Japanese creativity is like growing rice. Whereas wheat farming lends itself to large-scale, highly mechanized management and spontaneous growth, Japanese rice farming is an arduous, time-consuming process of cultivation.

The field must be prepared and flooded with water; the seeds are raised in nursery beds, then transplanted to the main paddy field in orderly rows. The field must be constantly monitored, weeded, and controlled to obtain the maximum yield. The rice farmer is literally at the beck and call of the paddy field.

The idea of cultivation (*ikusei*) is perhaps one of the more distinctive features of Japanese creativity. Whereas Western creativity is spontaneous, Japanese creativity is more studied and deliberate and is maximized by cultivating one's minimal resources. Large Japanese companies, for example, "cultivate" their employees by training them, sending them abroad, rotating them, and providing long-term financial support for new product development teams. They also allow a certain amount of play to encourage individual creativity within the context of group goals. Large U.S. companies also cultivate employees to some extent, but their shorter time horizon and higher employee turnover discourage them from investing as much time and money in employee training. Keisuke Yawata, President of LSI Logic K.K., reflects upon his days at NEC:

> In Japan, employees are tightly squeezed in all aspects of their lives. They have very little freedom like Americans. However, in NEC, employees were allowed to pursue their interests in areas beneficial to the company, which they did with great zeal. It was a kind of focused creativity.

The principle of *ikusei* underlies the Japanese government's policies of protecting and promoting infant industries, such as software, biotechnology, and new materials. The government establishes national research projects, sets collaborative goals, and allocates research fields among participating companies, which are provided with financial assistance, special depreciation allowances, and tax incentives. Helping the participating companies to succeed is a major goal of government. U.S. industrial policies, on the other hand, are much less structured and coordinated, more laissez-faire. We plant seeds, but do not carefully cultivate our crops. Neither do we "pick" winners and losers among our companies, preferring to let them succeed or fail on their own.

Thus, Western creativity tends to be spontaneous and overflowing—like a field of wheat or wildflowers—while Japanese creativity is compact, manicured, and highly structured—like a rice paddy or a bonsai plant.

Creative Fission vs. Creative Fusion

If cultivation is the mother of Japanese creativity, teamwork is its father. The goal of Japanese creativity is not just to create new products and ideas, but to also build teamwork and a sense of harmony. Japanese creativity thrives on group interaction and brainstorming. Individuals are encouraged to contribute ideas for the benefit of the team, not to be overly spontaneous or different. People who refuse to cooperate or continue to advocate unique ideas are distrusted, laughed at, or ignored by their peers in Japan. Many Westerners who believe the Japanese lack creativity do not see that the Japanese prefer to hide their ideas behind a facade of conformity to avoid being embarrassed in public.

As shown in Figure 2–1, Western creativity is like nuclear

FIGURE 2–1
U.S. vs. Japanese Modes of Creativity

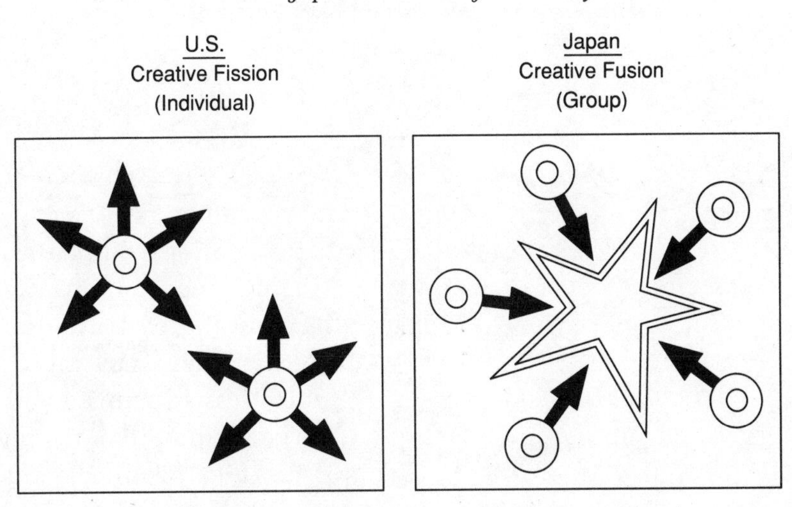

fission, in which individual brilliance has an opportunity to shine, while Japanese creativity is like nuclear *fusion*—ideas from many people are gathered, assimilated, and squeezed into a new product. Such a fusion of ideas—known as *yugo-ka* in Japan—can trigger an intense explosion of new ideas and is extremely effective in creating new products.

Katsuhide Hirai, Director of Information Systems at Fujitsu America, explains how group creativity works at Fujitsu (see Figure 2–2):

> The TQC [total quality circle] movement in Japan brought out creativity from ordinary people. Everyone contributes new ideas to the overall group, but nobody is willing to take credit for the idea because of our emphasis on group harmony. Showing off is looked down upon in Japan, so our star performers either quit or go to the university where they are freer to do what they want. In the United States, your stars usually grab the credit. Our creativity occurs underwater out of sight. Creative ideas from individuals are merged with those of the group, which are combined with good ideas from other groups to develop new products. On the surface, the product may not look creative or

FIGURE 2–2
The Japanese Approach to Creative Research

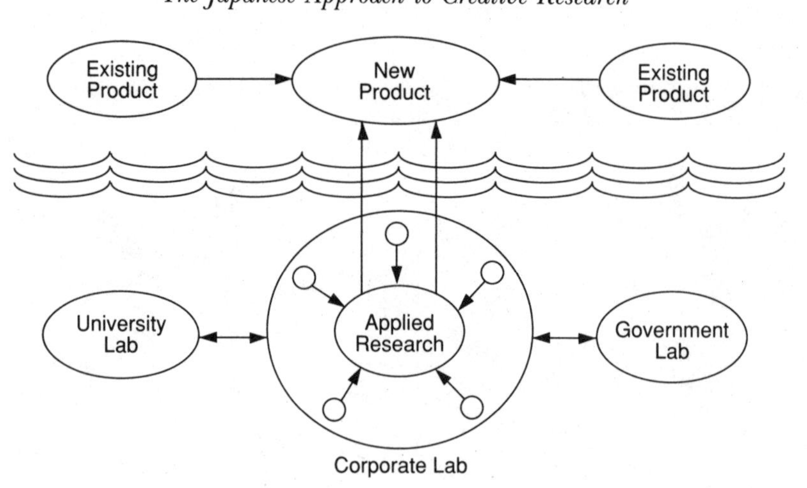

> unique to Americans, but it represents the best ideas of many people. The ideas are built into the product.

The impact of these contrasting approaches is well known. Mr. Hirai observes, for example, that American software programmers develop excellent software packages such as Wordstar or Lotus 1-2-3, which are sold individually, while Japanese programmers are much better at developing integrated software for use on large-scale in-house systems. Fujitsu spent five years developing computer-integrated manufacturing (CIM) software and systems for its Japanese plants—a project that involved over 100 people from dozens of Fujitsu companies—but the software will not be sold. Many Americans assume that Japanese are weak in software because they sell few software packages in the United States. But Mr. Hirai points out that Japanese video game software is very sophisticated and creative. He believes that Fujitsu's in-house software is as good or better than Western CIM software, which usually requires complex band-aid approaches to link totally incompatible systems.

He notes:

> Westerners are very creative in specialized fields, but their software is often incompatible and programmers dislike having to debug their software, so they end up with quality problems. But we Japanese take the time to fully understand the whole problem, then develop overall solutions. Our programmers are willing to debug their programs and work with the end users to refine it. This is probably the biggest difference between East and West.

Cartesian Logic vs. Fuzzy Logic

Perhaps one of the most frustrating aspects of the Japanese approach for Westerners is their circuitous mode of thinking. Whereas Western logic, reflecting our Cartesian heritage, follows a clear, linear path of reasoning, Japanese logic, largely influenced by Buddhism, is often "fuzzy" and cyclical. Japanese speakers frequently puzzle foreigners by "beating around the bush" before getting to their main point. Westerners tend to be more straightforward and frank, especially North Americans. As discussed in Chapter 4, Japanese think in circles, while Westerners think in straight lines.

The Japanese tolerance for vagueness also sets them apart from Westerners. While Americans tend to think in black-and-white (yes/no, we/they, against/for), the Japanese are much more comfortable with shades of gray. Indeed, it is often said that Western thinking is digital and Japanese thinking is analog. For the Japanese, problems are rarely clear-cut, with simple answers. Rather, everything is relative and open-ended. Problems are viewed as complex, incoherent, and vague, requiring incremental, diffused solutions rather than the hard-hitting, close-ended solutions commonly proposed in the West.

Uni-functional vs. Multi-functional Creativity

Because of their limited space and resources, Japanese constantly seek multiple applications for new ideas. A familiar example is the use of multi-functional rooms in Japanese homes; living rooms are transformed into bedrooms at night by rearranging a few sliding doors and pieces of furniture.

By contrast, Western rooms and furniture generally serve one purpose only. Similarly, Western creativity tends to be uni-functional—ideas are developed with a single use in mind. Now, however, the multi-functionality of Japanese architecture is beginning to be copied in the West as a result of rising land prices. For example, California builders of starter homes are combining the family room with the often unused living room and are using sliding glass doors that open onto small gardens to provide a feeling of spaciousness.

Multi-functional creativity has many applications. The Japanese transformer toy that can be converted from a race car into a Godzilla-like monster is a familiar example. On a larger scale, Japanese carmakers such as Mazda and Toyota are now using multi-functional parts to develop "design-your-own" cars. In the future, prospective car buyers will be able to choose from a variety of car chassis and body paneling units to custom-build their own cars. The human imagination is the only limit on the variety of multi-functional products that can be developed using common subsystems and components.

Creative Fusion: East Meets West

In early 1988, two young Japanese programmers from ASCII Corporation, a fast-growing software company, sat down at a Silicon Valley sushi bar. They were visiting their local joint venture partner. In 1986, ASCII, Kyocera, and Mitsui and Company had teamed up with Chips and Technologies, a Silicon Valley semiconductor start-up company, to develop "chip sets"—device packages integrating a variety of logic and memory chips—for IBM PC clone makers in Asia. Their alliance was a novel arrangement, and at the time, nobody took them seriously. Chips' president, Gordon Campbell, had come from a floundering chip company, and the three Japanese companies had no experience in developing chip sets. In the middle of a deep industry recession, their prospects seemed dim. Why had they even entered the joint venture? And what did the Japanese companies have to offer?

Ryoichi Kurata, ASCII's product marketing director, summed up the Japanese role in the partnership: "We're developing software to help Chips get into the Japanese market. So far, things are going well. We seem to complement each other. Americans are very good at creating new chip designs and software concepts, while we are handling the applications software. That is a Japanese strength."

However unlikely a union, their partnership has borne fruit. Since its founding in 1985, Chips and Technologies' revenues zoomed to $210 million in 1988, far surpassing the expectations of most industry watchers. Overnight, Chips single-handedly created a new chip market, giving IBM a run for its money from Asian PC clone makers. But more interesting is the powerful cultural synergy at work between this U.S. start-up company and its Japanese partners. Unlike larger U.S. semiconductor companies, which are battling the Japanese for survival, Chips has discovered and assimilated Japanese creativity.

During the 1980s, other Japanese and American companies also discovered the power of creative fusion. In 1988 Dataquest estimated that Japanese chip makers had teamed up with 142 partners, of which 100 were American or European. Unlike the technology giveaways of the past, these partnerships involve complex ventures, mutual product-sourcing, collaborative research, and technology-sharing. Toshiba, for example, is trading its memory chip technology and production know-how for Motorola's microprocessors through a $280 million joint venture, Tohoku Semiconductor, which is located north of Tokyo. Sun Microsystems, which hit $1 billion in annual sales after only six years, developed its SPARC microprocessor chip with Fujitsu. NEC, Siemens, and MIPS Computer are developing RISC (reduced instruction set controller) systems for global markets. Hitachi and VLSI Technology are sharing semicustom chip know-how. Intel is developing microprocessors for Canon auto-focus cameras and Nissan's new engine control systems. And Advanced Micro Devices (AMD) is jointly developing chips with Sony.

Today, yin is meeting yang in the corporate laboratory—and the resulting synergy is shaking the global economy.

Creative Global Fusion

Since the nineteenth century, Japan has looked to the West for new ideas and scientific knowledge. But because of the rapid growth in the Asian nations, Japan will turn more and more to its neighbors for creative new ideas. Former Prime Minister Noboru Takeshita signed an agreement with mainland China to discuss the long-term prospects for joint economic development and trade. The two nations are perfect complements; Japan has the financial and technical resources, while China has the land, people, and markets. Working together, they will develop new approaches to science and technology that have not been seen in the West.

In the past, economic dominance depended upon a country's creative strengths. But global markets will force competitors to adopt other forms of creativity. The most successful companies and nations will be those that are willing to learn from others, and creative fusion will become a strategic tool in an era of global competition. The Japanese seem willing to make the long-term investments required to achieve that goal.

3

The Creative Samurai: Japanese Intrapreneurs and Small Businesses

> . . .What were the ingredients of an excellent strategist, and how could they be reproduced in a Japanese context?
>
> The answer I came up with involved the formation within the corporation of a group of young "samurais" who would play a dual role. On the one hand, they would function as real strategists, giving free rein to their imagination and entrepreneurial flair in order to come up with bold and innovative strategic ideas. On the other hand, they would serve as staff analysts, testing out, digesting, and assigning priorities to the ideas, and providing staff assistance to line managers in implementing the approved strategies.
>
> —Kenichi Ohmae
> *The Mind of the Strategist*

IN the fall of 1980, general manager Mikio Naya of Minolta Camera Company looked at the stagnant 35mm camera industry and decided that Minolta needed to take a different tack to revive its competitiveness. Instead of merely improving on existing models, he wanted to create a totally new camera that would catch the industry by surprise. Minolta had a reputation for making highly sophisticated cameras, but its sales remained stagnant because its cameras were so difficult to use. If his team could develop an entirely automatic-focusing, single-lens reflex camera, Naya was sure Minolta could regain market share and, more important, its sense of confidence in the hotly competitive world of Japanese photography equipment.

To his dismay, he encountered strong resistance from his sales and engineering divisions because his proposal would require modifying the lens mount, a change that might turn away customers. Moreover, Minolta's decision-by-consensus approach meant that his idea could easily be shelved or torpedoed. But Naya was backed by a powerful ally, Minolta's president, Hideo Tashima, who had launched a companywide campaign to stimulate the development of creative new products.

Tashima explained:

> We had long considered how to move to automatic focusing, the great innovation in the industry. But since we weren't the industry leader, a cautious approach would expose us to the danger of dwindling sales. Consequently, we decided to gamble. In our eyes, the future of Minolta depended on taking a large risk and going beyond the others to create the camera for the twenty-first century.

Thus, in the face of strong internal dissension, Naya was given the go-ahead.

Minolta was the last of Japan's camera makers to introduce an auto-focus camera, which first appeared in 1981. The other companies had tried to crack the market, but they had designed the focusing motor into the interchangeable lenses, making the cameras too bulky and heavy for the average user. Naya proposed reducing the size of the lens by building the focusing motor into the body.

Although his idea was intriguing, Japanese electronics makers were reluctant to commit themselves to the project. They could not guarantee that they could develop the microprocessors or the automatic-focusing sensor required for the camera.

Tokuji Ishida of Minolta's Research and Development Department recalls:

> When the charge-coupled devices (CCDs) appeared back in 1981, we were really excited. We realized that CCDs would make fully automatic focusing possible. The problem, though, was price. If CCDs were too expensive, commercialization would be out of the question. I guess you could say that the development of the Alpha 7000 began with our repeated visits to the integrated circuit manufacturers, asking them to lower the price of their CCDs.

Despite the cost barrier, Naya pressed on. Working overtime for more than two years, his team came up with an entirely automatic single-lens reflex camera, the Alpha 7000. The camera featured a CCD sensor linked to two microprocessor chips that could instantaneously calculate the distance to the image and automatically focus the lens, and a family of twelve slim, lightweight lenses.

In 1984 Minolta had captured only 5 percent of the Japanese single-lens/reflex camera market and its corporate image was still languishing. In February 1985 the Alpha 7000 camera was introduced with a worldwide advertising blitz; in North America it was called the Maxxum 7000. To the astonishment of industry observers, the Alpha 7000 took off in the marketplace. Demand for the camera overwhelmed supply, forcing Minolta to double its monthly production to 60,000 units. Camera stores and distributors were backlogged with orders. By the end of the year, the Alpha 7000 had totally transformed Minolta's image: In December the Alpha 7000 was named to *Fortune* magazine's outstanding products list, and in Japan it received the prestigious Nikkei Award for Creative Excellence.

Minolta rapidly gained market share, leaping from 5 percent of the single-lens reflex market in 1984 to 26 percent in 1985. Company sales soared 47 percent in September 1985 alone because of the Alpha 7000, which has become the most successful auto-focus camera in Japanese marketing history.

The Alpha 7000 revived Japan's moribund single-lens reflex camera market, which had declined from 7.6 million units in 1980 to 4.9 million in 1984. Before the Alpha 7000 was released, the Japanese camera market was saturated with look-alike cameras and camera makers were diversifying into other businesses. After the Alpha 7000 appeared, Japanese camera production grew by 11.5 percent to 5.5 million units. A tidal wave of new product ideas and technological creativity was unleashed. Canon, Nikon, Olympus, Pentax, and other Japanese camera makers jumped onto the bandwagon, introducing stylishly designed 35mm cameras featuring zoom lenses, push-button controls, infrared focusing beams, and handgrips loaded with electronics. In 1988 the Chinon Genesis, the Olympus Infinity SuperZoom 300, and the Yashica Samurai battled for market share with the best-selling Maxxum 7000.

Minolta has already come up with more new ideas—such as programmable plug-in cards that control the Maxxum 7000's focusing, exposure, and film-advance systems for use in high-speed sports photography and other tough assignments. Its next set of innovations are in the works, and Japanese camera makers will undoubtedly be drawn into a fresh round of competition. Indeed, they have already learned a valuable lesson that will steel them against the onslaught of tough Asian competition—the benefits of giving free rein to the imagination and creative excellence of intrapreneurs like Mikio Naya.

Throughout Japan, companies are exploring new ways to become more creative: restructuring, opening new research centers, diversifying into new markets and technologies, running creativity training seminars, and turning their quality circles into "creativity circles." Still, the Japanese face innumerable social and organizational barriers to creativity. How are companies mobilizing to overcome these obstacles? What approaches are they using? Let's look at how Japan's large corporations and burgeoning service businesses are coping with the growing demand for creativity (see Figure 3–1).

FIGURE 3-1
Japan's Corporate Restructuring

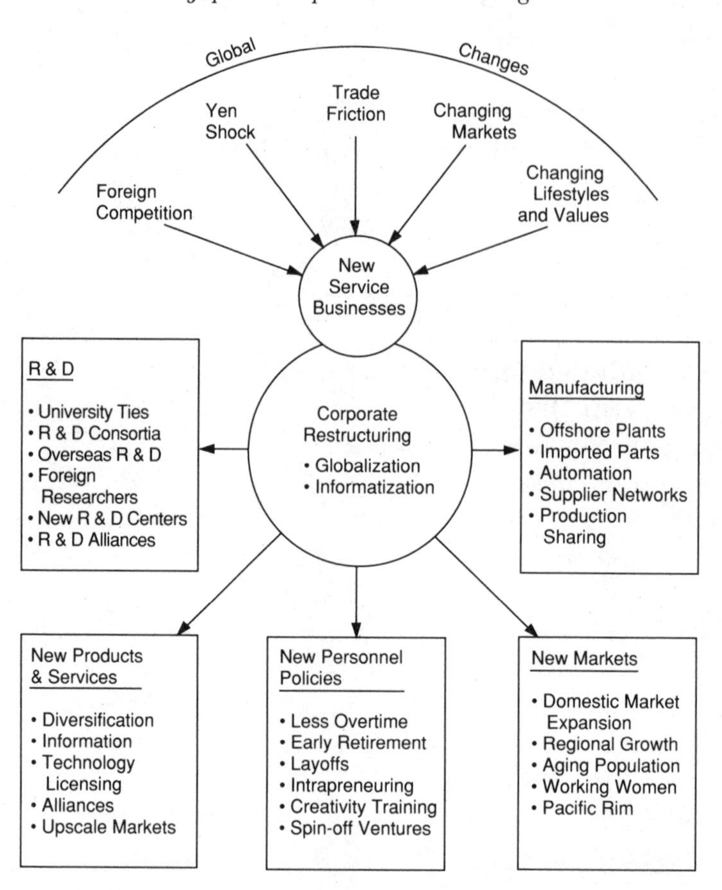

Streamlining Heavy Industries

In February 1987 Nippon Steel Corporation, the world's largest steelmaker, announced plans to lay off 19,000 of its 64,000 employees and close five blast furnaces by 1990. Coming at a time of major retrenchment programs by other steelmakers, this announcement sent shock waves throughout the Japanese economy. During the next few years, Japan's top steelmakers—Nippon Steel, Nippon Kokan, Sumitomo Industries, Kawasaki Steel, and Kobe Steel—plan to lay off 100,000 workers. And this pattern is being repeated throughout the country. Employment at Ishikawajima-Harima Heavy Industries (IHI), Japan's second largest shipbuilder, has dropped to 17,000 from 37,000 in 1972. Hitachi Shipbuilding Co. cut its shipbuilding work force from 17,500 to 5,800 in late 1985. From chemicals and railways to shipbuilding and aluminum smelting, Japan's heavy industries are laying off people and slashing costs in order to survive. Like their European and U.S. counterparts, heavy industries in Japan are struggling financially. The old Japan Inc. is sinking fast.

To boost sales in spite of downsizing and cost containment, these industries are also rapidly diversifying into new venture businesses. Nippon Steel is producing electronic parts in a joint venture with Philips of Holland, manufacturing silicon wafers for semiconductors, and marketing imported U.S. superminicomputers from Concurrent Computer. By 1995, steel will account for only half of Nippon Steel's sales; 20 percent will come from electronics and communications, and 10 percent each from regional development projects, services, and new materials. Like other steelmakers, Nippon Steel is moving downstream toward the assembly and marketing of high-tech products jointly developed between foreign firms and its own laboratories.

The Overseas Challenge

Steelmaker Nippon Kokan (NKK) is also diversifying as fast as it can. Since 1985 it has purchased polysilicon maker Great Western, beefed up its electronics research center, spun off new venture subsidiaries, and begun developing software for three U.S. superminicomputer makers selling into Japan—Masscomp of Massachusetts, Convex of Texas, and Ridge Computer of Silicon Valley.

Like other heavy industries, NKK faces numerous challenges in developing partnerships with foreign companies. The biggest hurdle is cultural, both corporate and social. Small, entrepreneurial U.S. companies can be puzzling to large Japanese corporations because of their rapid employee turnover, erratic direction, and lack of long-term planning. Trying to understand the mind-set of professionals from regions as diverse as Texas, Massachusetts, and California can be totally mystifying to the Japanese. Yoshimitsu Ido, research manager at one of NKK's new electronics divisions, represents a new breed of Japanese corporate intrapreneur learning to deal with Western culture. Educated at Tokyo University, he travels around the United States several times a year to develop NKK's ties with these small U.S. superminicomputer makers, working evenings and weekends to coordinate software development and Japanese marketing strategies.

Mr. Ido says NKK's high-tech diversification strategy often puts the company into an awkward situation. Many U.S. high-tech ventures see Japanese corporations as "bankers of last resort." He states:

> Recently, we have been approached by small, struggling U.S. high-tech ventures who want us to buy them out because they cannot get adequate financing from U.S. venture capitalists or banks. They hope to take their profits and start over or retire in Hawaii. In 1987, we were approached by over two dozen companies. I would like to work with U.S. engineers and entrepreneurs, but we turned most of them down because we didn't understand their business. We would rather work with a handful of companies than buy them out.

Less frequently articulated, but always present in the minds of many Japanese managers, is worry over a major U.S. political backlash and the highly emotional "Japan-bashing" that would result from buying out too many U.S. high-tech companies. Ever since the Pentagon's staunch opposition in 1987 to Fujitsu's attempted buyout of semiconductor maker Fairchild, Japanese companies have kept a lower profile. Nowadays Japanese companies often swap technologies and exchange engineers with U.S. companies and acquire minority ownership in smaller U.S. companies. Sony and Kyocera's investment in Vitelic, a Silicon Valley memory chip maker, is a case in point; Kyocera president Kazuo Inamori sits on the Vitelic board as a part investor.

Doing business overseas is only half of the Japanese intrapreneur's challenge. At home, cumbersome decisionmaking practices and layers of bureaucracy are anathema to corporate intrapreneurs like Mr. Ido. The *ringi* system—a consensus-building practice in which proposals are routed around the company for approval by managers related to the project—is not only time-consuming but stifles new ideas. Mr. Ido explains how his company loosened up this rigid system:

> NKK's president set up high-level business development groups that report directly to him. The groups only have three to five people—very creative people from research, manufacturing, product development, and marketing. We keep the groups small to encourage innovation. That's how we set up our electronics research center and subsidiary; we went around the organization.

Even if they overcome their own bureaucracies, Japanese companies are known for stifling initiative and creativity because of their emphasis on group harmony. In most Japanese companies, highly individualistic or creative employees are still considered oddballs or outsiders. Mitsuru Misawa, general manager of international investment at the Industrial Bank of Japan, believes that more and more Japanese companies are realizing how important it is to take the reins off creative individuals and to let them work up to their unique abilities: "In addition, young employees are now demanding interesting jobs which are more suited to their individuality. This attitude on the part of youth may lead to a new style of management emphasizing the importance of creativity."

The Internal Challenge

Kao Corporation, a $2.8 billion soap and detergent maker, is one of Japan's more innovative companies; it has a reputation for actively involving its employees in creative product development. President Yoshio Maruta is known as a "philosopher-manager" because of his studies of Confucianism, Buddhism, Christianity, and other religions. Applying the teachings of Shotoku Taishi (Japan's sixth-century ruler) and Zen, Maruta gives employees equal access to all company information because he believes that to encourage creativity, "knowledge and actual experience must become one." Professor Ikujiro Nonaka of Hitotsubashi University has observed:

> There are no top management–only secrets at Kao. Any employee can draw information freely from a companywide data base, attend meetings of the board of directors, and participate in monthly research and development conferences. . . . At Kao this is referred to as "experiential communication," and it plays a vital role in the new product conceptualization process.

Using this approach, Kao doubled its sales between 1977 and 1986 when it diversified out of the soap market into diapers, medical items, floppy disks, and chemical and synthetic products.

Kao Corporation's route to success is an exception rather than the rule. By and large, corporate diversification has not been a successful way to develop creative products and services and is not a cure-all for the Japanese economy. A Nippon Credit Bank study discovered that 70 percent of the new manufacturing ventures formed by twenty large corporations remained unprofitable. Likewise, a Nikkei Business survey of 809 diversification ventures formed between 1981 and 1986 found that 72 percent were unprofitable or went bankrupt within a few years.

Expanding Small Businesses

Despite their financial resources and ability to attract elite college graduates, large Japanese corporations generally lack the speed, flexibility, and creativity required to operate on the leading edge. Increasingly, as in the United States, Japanese small businesses are filling the void with their creative new products.

According to the Ministry of International Trade and Industry (MITI), small businesses provided 72 percent of Japanese jobs in 1985. While employment at large corporations declined from 3.31 million in 1975 to 3.07 million in 1985, small business hiring grew from 7.35 million to 7.81 million. MITI expects new small businesses to remain the fastest growing sector of the economy in the 1990s. The *Japan Economic Journal* reached similar conclusions in its 1987 survey of the top 1,000 small businesses, which had total sales of $24 billion in 1987. Their employment had increased 20 percent despite the 1985–86 recession caused by the yen shock. The small businesses surveyed were planning to expand their hiring by another 22 percent by 1990.

Traditionally, small businesses have been the most creative sector of the Japanese economy. In the postwar era, start-up companies such as Sony, Hanae Mori, Kyocera, and Sord Computer created new markets and industries, often in defiance of Japan's stifling business environment. The first "venture business" boom, which peaked with enthusiasm and optimism in the 1960s, ended with the 1973–74 oil crisis and a series of wild investment schemes. It was followed by a second venture boom in the early 1980s, a period that saw the rise of numerous high-tech companies—such as Cosmo AT, Arback Seihaku, Sodec, and GRAPHICA—and thirty new venture capital companies. But the yen shock, conservative venture capital practices, a lack of "cash-out" options, and a handful of spectacular bankruptcies have dampened the enthusiasm for venture start-ups.

Recently, there has been a third venture business boom because of Japan's rapidly growing economy and the shift toward value-added services. Japan's venture capital market jumped from $560 million in 1984 to $1.6 billion in 1987, and there are now over eighty venture capital firms. Although their fund-raising is based on the American approach, Japanese venture firms are primarily run by large financial institutions with little experience in managing small venture businesses. According to Yoriko Kishimoto of Japan Pacific Associates:

> They spend most of their time searching for and recruiting the very few promising young companies, and not enough time helping to set up a fruitful growth environment for them. There is still a lack of trust and communication between the investor and investee. From many entrepreneurs' point of view, venture capitalists come calling only after one's need for them is gone.

The IPO (initial public offering) market is beginning to improve: 125 venture capital–backed companies went public in 1986, but most received their venture funding just before going public. Over the next few years, Japan's tremendous capital asset liquidity, the demand for creative new companies, and the growing number of independent-minded young people will stimulate a growth in the venture capital market.

The Japanese government is getting more involved in encouraging small business. Traditionally biased toward large export-oriented corporations, MITI has developed programs to help small and medium-sized businesses become more innovative. MITI also offers the typical "infant industry" promotion measures—a 7 percent tax deduction and 30 percent special depreciation allowance on capital equipment purchased by small ventures.

The New Business Association, created in 1985 by over 200 leading industrialists, organizes marketing research, foreign tours, international symposiums, and technical exchanges for small businesses. Unlike MITI's technology-oriented Venture Enterprise Center, the New Business Association is trying to create new "hybrid innovation" services by introducing high-tech equipment into traditional businesses and providing training and personal networking. Family restaurants, for example, would receive help in installing electronic kitchen systems linked to distributors and service suppliers. Venture business associations have sprung up in Nagoya, Sendai, Fukuoka, Osaka, and other Japanese cities to promote entrepreneurialism. "Venture incubators" are extremely popular, especially outside of Tokyo.

Since 1988 MITI has encouraged small companies from different industries to pool—or fuse—their knowledge in order to develop new technologies and markets. MITI has developed the "technology fusion index," which shows the level of research spending by two different industries in similar fields, thus pinpointing the potential for synergistic research (see Chapter 5 for a more detailed discussion). Takayuki Matsuo, former deputy director of MITI's Small and Medium Enterprise Bureau and Stanford fellow, observes that "many small businesses are seeking new ways to use their older technologies. We are working with them to recycle their technologies using our 'fusion technology' method. Perhaps fusion—*yugo-ka*—is a form of Japanese creativity." At a recent Nagasaki Technopolis Symposium, for example, five companies announced an unmanned marine robot that was jointly developed for underwater resource exploration and aquaculture.

Today, over 20,000 small businesses throughout Japan are participating in MITI's Small Business Fusion Promotion Program. They are experimenting with a variety of fusion management techniques:

- Technology fusion (the combination of two older technologies to create a new hybrid technology, such as bioceramics and biocommunications)
- Technology-marketing fusion (collaborative efforts between researchers and marketing divisions)
- Technology-service fusion (the infusion of new technologies into traditional service industries)
- Marketing fusion (collaborative efforts by different product marketing groups)
- Finance-service fusion (the application of new financial techniques to existing services)

Whether the New Business Association and MITI's Small Business Fusion Promotion Program will succeed is still uncertain, but it is worth noting that MITI's regional testing laboratories have been very successful at transferring technology from regional laboratories to local businesses. In its role as policy coordinator, MITI provides valuable ideas, industry contacts, and the stamp of approval that legitimizes small businesses, which usually have difficulty securing adequate financing.

In the past, MITI has had mixed success: it succeeded brilliantly in helping the semiconductor industry, but played a negligible role in the success of the automobile industry. As Japanese companies respond to new demands for creativity, the Japanese government is being called upon to play the role of catalyzer and adviser rather than industrial coordinator.

Despite the emergence of corporate intrapreneuring and a venture capital market, developing creative businesses presents a formidable challenge to the Japanese. As effective as Japanese management has been in the past, many practices will be obsolete in the future. To overcome the mental roadblocks posed by its propensity for homogeneity, Japan will need many creative samurai like Minolta's Mikio Naya—intrapreneurs who are willing to risk their careers and their companies' future on new breakthrough ideas. If Japanese companies do not welcome such efforts, they will succumb to gradual decay and decline, and the "Japanese miracle" will be nothing more than a short-lived mirage. If the past is any guide, however, the Japanese are not likely to give up easily in their pursuit of creativity. As a Japanese proverb goes: "Knocked down seven times, back up eight."

PART II
Japanese Approaches to Creativity

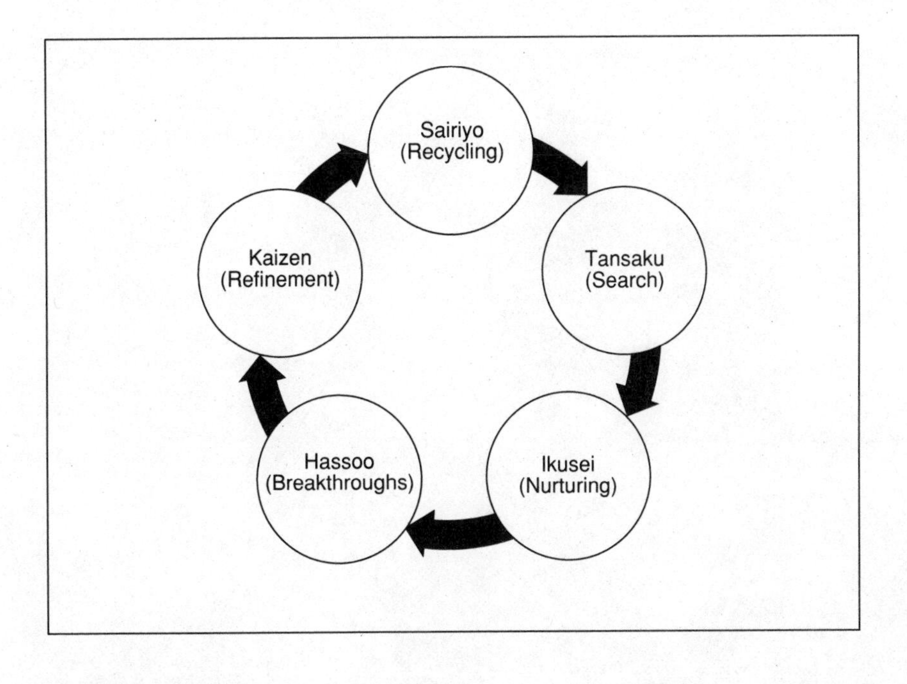

4

Historical Origins of Japanese Creativity

> Now we need true Japanese creativity and Japanese technology. . . . *Mukta* involves abandoning oneself and annihilating the ego to become "free," and allowing creativity to emerge.
>
> —Sueo Matsubara
> President of Automax

TODAY, a journey is quietly under way in Tokyo that promises to have a profound influence on the nature and direction of Japanese science and technology in the twenty-first century. It is an inward journey into the heart of Japanese culture and philosophy, a search for "paradise lost": the origins of Japanese creativity in the values shared by every Japanese. Unlike in the past, travelers on this journey do not rely on government-sponsored projects or foreign inspiration. By looking inward and grounding themselves in the bedrock of Japanese traditions, they seek to break through the psychological barriers preventing the Japanese from fully tapping their creativity.

Exploring the Buddhist Roots of Creativity

In the Hanzomon district, tucked away in a quiet office and residential area not far from the Imperial Palace, is the Mukta Institute, an oasis in Tokyo's commercial hustle and bustle. At first sight, the institute is not very impressive. The small meeting room—cluttered with bookshelves, an electronic blackboard, and a yin-yang–shaped table placed askew on the tatami mats—looks more like a converted tea ceremony room than a place of serious thinking. But in this room is the nucleus of Japan's creative leadership. Its members are a Who's Who of Japanese industry—Soichiro Honda of Honda Motors, Kazuma Tateishi of Tateishi Electric, and Masahiro Mori, Tokyo Engineering University professor and founder of the Robotics Society of Japan.

Founded in 1970 by Masahiro Mori, the Mukta Institute, or *Jizai Kenkyujo* as it is called in Japanese, is an alternative-style think tank that merges scientific rationality with Buddhist philosophy to arrive at creative new solutions. It is the heir to Japan's historical search for a satisfying union of Western science and Eastern philosophy (*wakon yosai*). The Sanskrit term *mukta,* like its Buddhist equivalent *jizai,* refers to the personal freedom achieved through spiritual enlightenment, attained by dropping old preconceptions, abandoning oneself, and in Zen fashion, eliminating the ego to become free and creative. At its core is Rinzai Zen, which uses paradoxes and riddles to shatter rigid thinking, thus allowing personal truths to emerge.

The Mukta Institute has over forty core members who regularly meet to develop creative insights by reciting Buddhist scriptures, meditating, and discussing difficult problems. Membership is open to the public, which is encouraged to participate in institute study groups. The Mukta Institute gospel is Mori's unorthodox book, *The Buddha in the Robot: A Robot Engineer's Thoughts on Science and Religion,* in which Mori argues that "robots have the buddha-nature with them—that is, the potential for attaining buddhahood." His use of robots is a metaphor for the buddha-nature in Japanese society and, ultimately, for its values as expressed in high technology. In his recent book *Advice on Nonduality* (*Mufunbetsu no Susume*), Mori argues that conventional wisdom and judgment obscure reality and that the Japanese need to follow the path of nonconventional wisdom if they wish to release their creative powers.

Sueo Matsubara, acting president of the Mukta Institute, approaches his work with a distinctive flair. A jovial man with an infectious grin, he gesticulates with his hands and body while he speaks. Leaping up from the tatami, he points to the volumes of Buddhist scriptures lining the shelves:

> If you really want to understand the essence of Japanese creativity, you must study Buddhism. It is the key to Japanese thinking and culture. You can either study the scriptures of Pure Land Buddhism or practice Zen meditation. We call these approaches *tariki-hongan* [salvation through the benevolence of Buddha] and *jiriki* [salvation through one's own effort]. In either case, it's better than wasting your time reading business books and how-to manuals. Creativity cannot be taught mechanically—like golf. It's something that comes from within.

The Mukta Institute has clearly left its imprint on the business decisions of its members. Soichiro Honda, for example, conducts an annual idea contest at Honda Motors to elicit creativity from his employees, a contest that Toyota Motors has copied with its "Idea Olympics." Kazuma Tateishi sponsors the "Omoro Festival" to promote unique new products; they may not be profitable, but the festival stimulates his employees to think creatively.

Matsubara's own study of Buddhism has paid off for his robotics company, Automax. Proudly showing a videotape of his company's latest inventions, he bubbles with enthusiasm over his new robots. One is impressed by their variety and originality. Automax has developed a self-steering floor-cleaning robot, an underwater robot submarine, a merchandise display robot, a robot for removing sludge from inside crude oil tanks, car test-drive simulators, and dancing toy robots that were exhibited at Tsukuba Expo 85.

Business matters aside, Matsubara believes that the Mukta Institute is pursuing research critical to Japan's high-tech future.

As Japan becomes a technological superpower, he worries that it could go down the path of "soulless" science.

> As we have seen in history, science without religion is dangerous. Religion, especially peaceful religions like Buddhism, must be at the heart of science. For researchers and scientists, God must be within one's heart, not outside. That is what we are trying to accomplish at the Mukta Institute.

The Mukta Institute members are a tiny minority in Japan, a hardy band of pioneers venturing into the core of Japanese culture to understand the dynamics of Japanese creativity. In many ways, they are not unlike the transcendentalists of nineteenth-century Boston, who defied conventional wisdom by seeking the great "oversoul" in nature. By returning to their cultural roots, Professor Mori and his followers are exploring the frontiers of Japanese creativity. They are going where only Buddhist priests and Zen monks have trod before, in hopes of elucidating the principles that will guide Japanese researchers in the future.

The Mukta Institute is still ahead of its time. Most Japanese lack the time or interest for mastering esoteric Buddhist texts. Increasingly, they would rather spend their time eating and drinking with friends, reading, painting, writing, listening to music, relaxing, playing sports, unwinding at country spas, or traveling abroad. Leisure activities will continue to occupy the minds of most Japanese, but many will soon tire of hedonistic pursuits. As the baby boomers age, many of them are beginning to ask: Is there more to life than material wealth? How do I fit into society? How can I become a better, more well-rounded person? They are seeking a more spiritual, family-oriented lifestyle. Already, Buddhist study groups are forming at the rate of one every week in Japan.

Long commutes, marathon working hours, high prices, and cramped housing are taking their toll on Japanese office workers, who are suffering from frustration and stress. Worse, finding the time and space to pursue leisure activities is so difficult in Japan that many people seek solace in alcohol. As an alternative, many companies offer stress-reducing "meditation rooms" and sponsor back-to-nature retreats as a respite from the daily grind and to promote the emotional well-being and creativity of their employees. Zen is the simpler path.

Seeking Creativity in Zen

Probably no other mode of Japanese thinking contrasts more sharply with Western thought than Zen meditation. Indeed, Zen lies at the core of Japanese creativity because of the natural spontaneity and fresh insights it makes possible. A comparison of Western and Japanese thinking highlights the differences.

In the West, we tend to value rational, scientific thought over intuitive, "soft-headed" thinking. From childhood we are taught the Socratic method of inquiry: pursuing a series of rationally framed questions and answers. Along with public debate and discussion, intense cross-examining is our primary tool for scientific exploration.

In Japan, the closest historic parallel to the Socratic method is the *Zen-mondo* (question-and-answer session) in which a *koan,* or rationally unsolvable riddle, is posed by the Zen master. Unlike the Socratic purpose, there is no rational solution because the goal is spiritual enlightenment (*satori*) and intuitive understanding. Zen Buddhist philosophy does not place great value on rational thought alone. Daisetz Suzuki, professor of Buddhist philosophy at Kyoto's Otani University, explains:

> *Prajna* (intuition) is the self-knowledge of the whole in contrast to *vijnana* (reason), which busies itself with parts. *Prajna* is an integrating principle, while *vijnana* always analyzes. *Vijnana* cannot work without having *prajna* behind it; parts are parts of the whole; parts never exist by themselves, for if they did they would not be parts—they would cease to exist.

This emphasis on intuitive understanding explains in part Japan's traditional weakness in basic scientific research, in which logical reasoning plays a central role. While Europe's leading minds were pursuing theoretical research during the Renaissance, Japan's greatest minds were focused on politics, commerce, and spiritual enlightenment. Writer Hajime Nakamura observes among Japanese "an absence of theoretical or systematic thinking, along with an emphasis upon an aesthetic and intuitive and concrete, rather than a strictly logical, orientation."

During the Ashikaga period (1338–1573), Japanese sensibilities were heavily influenced by Zen Buddhism, and a strong cultural tradition emphasizing form, etiquette, and design gradually evolved. The tea ceremony (*cha-no-yu*), the use of natural woods and settings in architecture, rock gardens, Noh theater, and brush painting (*suiboku*) all reflect the discipline and simplicity of Zen. These are all creative outlets for expressing the inexpressible tenets of Zen, channels for the deep feelings and insights brought to the surface through meditation.

The Zen heritage explains why Japanese creativity is rooted in the design arts, which depend on synthesis and holistic, "right-brain" thinking, as opposed to rational, "left-brain" analysis. The Zen taste for refined simplicity and quiet (*wabi*), combined with a predilection for visual design, can be discerned in many Japanese products, ranging from televisions and stereos to modern architecture and fashion design. The Japanese demonstrate in many of their endeavors an uncanny ability to pull together disparate ideas and concepts in a continual process of visual refinement and synthesis. This ability is the Zen of Japanese creativity.

Because of the nonintellectual nature of Zen, however, Japan's base of scientific knowledge and research methodologies never developed as fully as in the West. While many Western scientific discoveries have involved intuitive leaps of imagination, these breakthroughs are grounded in scientific research. Relying on intuition and spiritual enlightenment alone makes for bad science. This scientific weakness has been a major handicap for Japan, which is seeking new models of creativity to complement or replace Western ones. As we'll discuss later in this chapter, the Japanese are searching for a way to integrate Western science and Japanese philosophy—one of the holy grails long sought by Japanese thinkers.

Thus, even though most Japanese are not practitioners of Zen, they are deeply influenced by Buddhist philosophy and the deeply held values it embodies. Zen will be one of the touchstones of Japanese creativity in the twenty-first century.

The acting president of the Mukta Research Institute, Sueo Matsubara, believes that fostering scientific creativity will not come easily to the Japanese because of deep philosophical rifts that must be reconciled within themselves. While Buddhism may offer a path of Japanese creativity, Confucianism, Japan's other legacy, is a two-edged sword: as a way of thinking, it has readied the Japanese for Western science, but has stunted individual creativity.

Michio Morishima, an economist affiliated with the London School of Economics, writes in his book *Why Has Japan Succeeded?*:

> In Japan today, Confucianism influences the everyday conduct of Buddhists and Shintoists and even Christians. Confucianism, which the Japanese view as an ethic rather than a religion, holds the following virtues as most important: loyalty to the state or the emperor, filial piety, faith in friendship, and respect for elders. It is primarily concerned with the individual's relationship to various communities. Confucianism discourages individualism. It is intellectual and rational in character, rejecting the mysticism and incantation common to other religions. The ability of the Japanese to assimilate Western technology and science with astonishing rapidity after the Meiji Restoration was due, at least in part, to their education under Confucianism; Western rationalist thinking was not entirely foreign.

Yet, because of its reverence for hierarchy and order, Confucianism is a barrier to change. Since it has served Japan so well in the past, the Japanese are unlikely to drop it entirely for a Buddhist view of the world. Indeed, Confucianism is likely to remain a powerful force behind Japanese educational and scientific policies, especially as other Asian nations with Confucian backgrounds—South Korea, Taiwan, Singapore, and China—challenge Japan in high-technology fields. Confucianism has nurtured the conservative nature of Japanese society and has formed the basis for Japanese "techno-nationalism"—or the hoarding of technology. By contrast, a Buddhist approach appeals to those attracted to the notion of sharing technology—or "techno-globalism."

Overcoming the Tyranny of Western Creativity

In the 1990s, Japan's key challenge will be to overcome the cultural domination of the West. The Japanese must dispel the notion that individual creativity is everything and develop their own blend of individual and group creativity.

An important landmark in scientific research was Thomas Kuhn's 1982 publication, *The Structure of Scientific Revolutions.* Kuhn challenged the prevailing view that scientific progress is a simple process of accumulating discoveries and knowledge. He argued that discoveries only have meaning within a "paradigm," or worldview, which is based on common assumptions about the problems being considered. Like one's cultural background, paradigms are a useful shorthand for understanding and interpreting the world. Usually implicit and unarticulated, they are "road maps" that influence the way people perceive, think about, analyze, and respond to their environment.

As long as the underlying assumptions and environment remain unchanged, paradigms are useful tools for analyzing and solving problems. But if the world changes rapidly, as it is doing now, they can mislead people about the changing realities. Postwar assumptions about the backwardness of Soviet science, for example, prevented U.S. policymakers from anticipating Sputnik. Likewise, global leaders were caught off guard by the 1973 oil shock because they had assumed that Western corporations would continue to control Middle Eastern oil fields. In both cases, there was ample evidence that the underlying assumptions were no longer valid. News of the scientific, economic, and political events leading to Sputnik and the oil shock was readily available, but most people ignored the impending signs because they conflicted with prevailing beliefs. Not until the events occurred were the "paradigms" suddenly forced to change.

Our view of creativity follows a similar pattern. In the West, creativity is viewed as an epiphany, and only one phase in the creative process—the generation of new ideas that trigger dramatic breakthroughs—is emphasized. In contrast to the incrementalism of Japan's adaptive creativity, Western creativity is rapid-fire, awe-inspiring, and often engenders the zeal of religious faith. Westerners delight in spectacular displays of genius by individuals. Despite the months or years of tedious groundwork, we believe that breakthroughs are the ultimate proof of creativity. Anything short of spectacular quantum leaps is considered "ho-hum" science.

While Western creativity has unleashed many new ideas, its rational bias has led us into many political and economic dead-ends. No longer are "big" solutions adequate to solve the complex global problems that face us. We are like baseball players trying to win the game by hitting only home runs. Complex problems require creative problem-solving. But our "big breakthrough" mentality overshadows subtler, less visible forms of creativity that are useful for reordering reality into new patterns.

Creativity can be evolutionary as well as revolutionary, convergent as well as divergent. It need not be limited to literary masterpieces or scientific breakthroughs. It can be holistic as well as specialized. In the broadest sense, creativity reflects a fresh, novel, and unorthodox way of thinking, living, and viewing the world. We need to expand our Western notion of creativity to include all forms of creativity, including Japanese creativity. What is this new worldview?

Mandala: A New Approach to Creativity

In the West, creativity is seen as a linear process based on individual effort. Roger von Oech, author of *A Kick in the Seat of the Pants,* articulates this view:

> I've found that the hallmark of creative people is their mental flexibility. Like race-car drivers who shift in and out of different gears depending on where they are on the course, creative people are able to shift in and out of different types of thinking depending on the needs of the situation.

Von Oech notes that people play four different roles in the creative process: the explorer (when searching for new information), the artist (when turning the information into new ideas), the judge (when evaluating the merits of an idea), and the warrior (when carrying the idea into action). He might have added a fifth role, often overlooked by Americans, but which the Japanese excel at: the antique dealer (when recycling old ideas for new applications).

By contrast, the more cyclical form of Japanese creativity can be viewed as a "mandala of creativity" divided into five related phases, as shown in Figure 4–1:

- Idea recycling (new uses for old and existing ideas)
- Idea exploration (the search for new ideas when existing ideas are inadequate)
- Idea cultivation (the seeding and incubation of new ideas)
- Idea generation (new breakthrough ideas)
- Idea refinement (improving and adapting new ideas to the changed environment)

FIGURE 4–1
The Mandala of Creativity

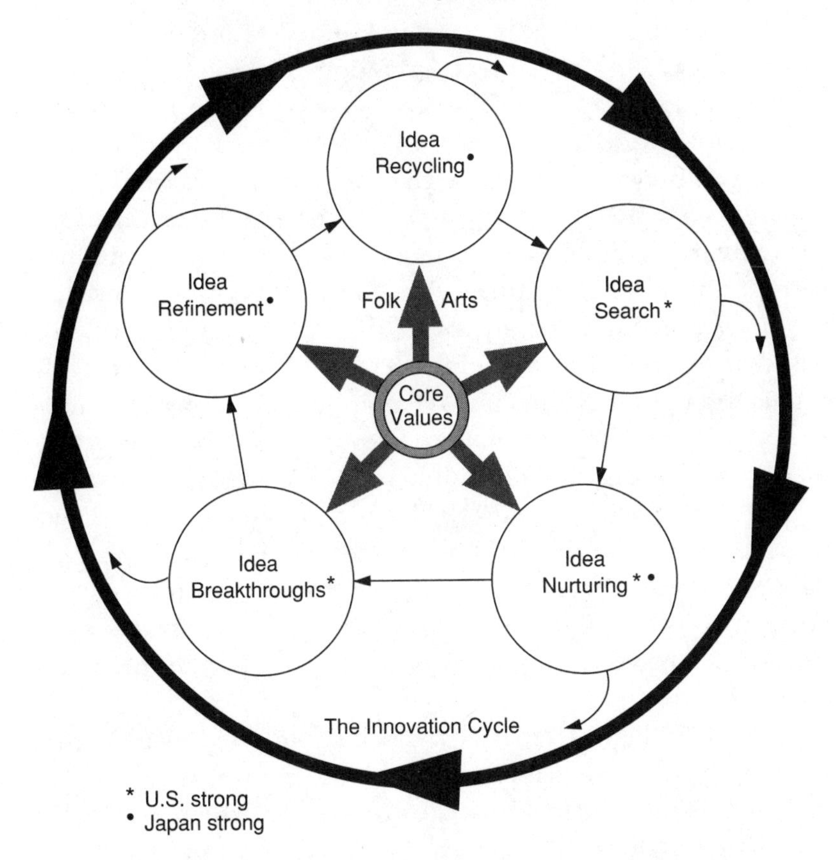

In this view, old ideas never die but, like energy, are transformed into new ideas as they race around the mandala. Although they may lie quietly for years or even centuries, a new development may trigger their reappearance in the world—a form of creative reincarnation. For example, many of Leonardo da Vinci's ideas and the science fiction of H. G. Wells could not be realized in their times because of limitations in existing technologies; now these ideas are technically possible. Their ideas about the helicopter and the submarine were far ahead of their times. As the Japanese have long known, it is worth reviewing and recycling old ideas from time to time because of their renewed potential.

Historically, the mandala has played a central, though understated, role in Japanese society. The mandala is a symmetrical, symbolic diagram used in Hinduism and esoteric Buddhism to express fundamental religious doctrines. Initially, it was a Hindu symbol meaning circle, group, or company. During the Heian period (794–1185), in Jodo (Pure Land) Buddhist scriptures, it symbolized supreme inner enlightenment (*satori*) and reflected the Buddhist view of existence as an unending cycle involving higher and higher levels of spiritual enlightenment with each succeeding cycle. Another type of mandala is the *suijaku mandara* (mandalas of the manifestation), which is based on the belief in the reincarnation of Buddha. During the Kamakura period, Amida mandalas appeared among Pure Land Buddhists who believed that repeating *Namu Amida Butsu* ("I place my faith in Amida Buddha") would bring them salvation. In the Muromachi period (1333–1568), thirteen-Buddha mandalas appeared among esoteric Buddhists for worshiping the deceased. Thus, throughout Japanese history, the mandala has been an object of contemplation for achieving the state of *satori*. It provides the cultural paradigm for Japan's pursuit of greater creativity.

The mandala of creativity can be conceived as a swirling solar system in which ideas unfold from central core values—depicted as the sun in many cultures—upon which a society is based. Core values such as nondualism (the union of man and nature, subject and object), purity (psychological, social, and biological), and adaptiveness are the centerpiece of Japanese culture. They are expressed in arts and crafts, the "spokes" of the mandala of creativity.

Japanese companies find that exploring creativity using the mandala structure is a dynamic, whirling process. They set up the traditional *gonin-gumi* (five-person teams), which work together to develop and refine new ideas. These groups resemble planets revolving around the sun; they go through all five phases of the mandala (see Chapter 8 for further discussion of these teams). Although group creativity predominates, individual creativity gradually emerges in this process. It resembles moons orbiting around each group, however, in contrast to the "lone genius" notion of Western creativity, which is more like meteors shooting through space.

From this swirling flow of ideas comes innovation, which is a concept very distinct from creativity. While creativity refers to pure ideas, innovation is the translation of ideas into tangible products and intangible services. Thus, the Japanese are very good at creative refinement (such as refining existing theories about video technology), as well as creative innovation (translating these theories into innovative new products). The distinction is subtle and often confused by Westerners. Not all creative ideas are innovative (or turned into tangible products and intangible services), nor are all innovative ideas necessarily creative—they may simply be well executed.

Sueo Matsubara of the Mukta Institute takes the mandala of creativity a step further.

> Creativity is not a one-dimensional circle. Like reincarnation, it is an unending process of refinement and recycling. The mandala of creativity is really three-dimensional. Each time you go around, the idea should get better and better. Otherwise, you're only going around in circles and going nowhere.

Thus, the Japanese notion of creativity can be visualized as a helix, in which each revolution through the cycle leads one to higher and higher levels of creativity (see Figure 4–2). The ultimate level of creativity, if it can be achieved, is *satori*, or spiritual enlightenment, in which the creator and idea become one. On a more philosophical level, the mandala of creativity is analogous to one strand in the double helix of DNA—the basic chain of life—or to the yin and yang of creation.

Western creativity is clearly stronger in three phases of the mandala process: idea exploration, idea cultivation, and idea generation. Westerners have traditionally excelled in pursuing basic research and exploring new scientific frontiers, activities that require maximum intellectual curiosity and adventurous-

FIGURE 4–2
The Evolution of Creative Japanese Ideas

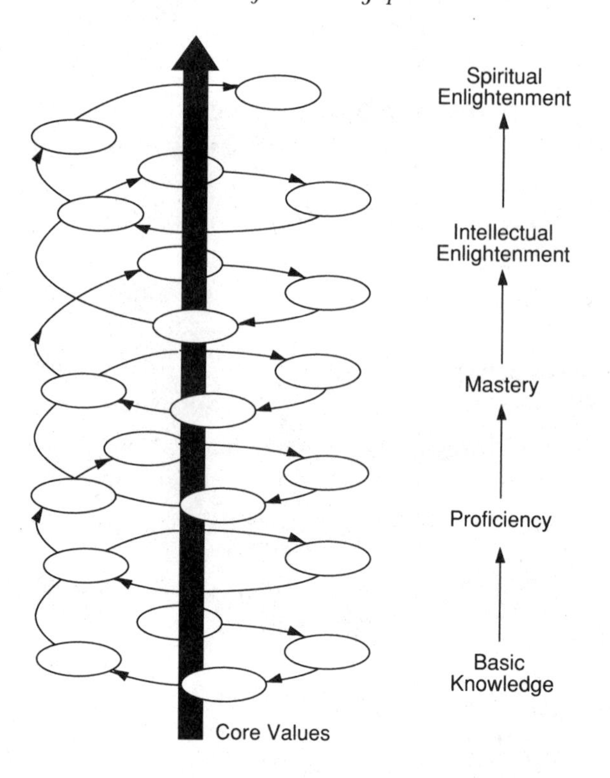

ness. By contrast, the Japanese, who have not been explorers or inventors as Westerners have, are strong in three other phases of the mandala: idea cultivation, idea refinement, and idea recycling.

Whereas the Japanese recognize their weaknesses—basic research (idea exploration) and breakthrough thinking (idea generation)—and are trying to correct them, Westerners mistakenly believe they have a monopoly on creativity and disregard the two areas where the Japanese are stronger—idea refinement and idea recycling. If this myopia continues, the Japanese could eventually master the entire mandala of creativity, which would have a devastating impact on Western industry in the twenty-first century.

But our mutual weaknesses need not amount to a zero-sum

game. Since Japan and the West have complementary strengths, both stand to gain from cooperative research. The Japanese bring a new way of viewing and using ideas that have been generated in the West. Conversely, Japanese approaches to creativity—such as miniaturization, technology fusion, and idea recycling—offer Westerners prospects for new creative breakthroughs. Indeed, yin needs yang to be whole. Growing techno-nationalism on both sides, however, presents an obstacle to the sharing of ideas. This is largely a political and industrial problem, not a technical or cultural one. Japanese and Westerners can work together well if given the opportunity.

From Folk Arts to High Technology

As the Japanese focus their creativity on basic research and technological breakthroughs, they will tap into a rich cultural heritage for guidance. They will apply many of their aesthetic principles and concepts—the spokes in the mandala of creativity—to problems of science and technology. This is already happening in corporate research laboratories throughout Japan. What are these aesthetic principles? And how are they being applied?

The concept of miniaturization is perhaps Japan's best-known form of aesthetic creativity, one that has permeated Japanese culture throughout history. Take, for example, miniature Japanese wood carving, or *netsuke*. In the Kamakura period (1185–1333), Japanese men wore small wooden hanging objects, or *sagemono,* to secure small purses to their sashes. These objects were adorned with intricately carved *netsuke* figures of Buddhas, wild animals, and natural scenes. By the Edo period (1600–1868), *netsuke* had risen to the level of high art and was worn by samurai and merchant classes alike. A symbol of personal wealth and creative taste, *netsuke* became so pervasive that gradually the Japanese became accustomed to the idea of carrying around miniature products.

Today, Japan's knack for creative miniaturization abounds

in the electronics world, ranging from Sony's miniature radios, Walkman audiocassettes, and pocket color TVs to its portable Diskman compact disk (CD) player. Sony is not the only creative miniaturist. Fujitsu, Hitachi, NEC, Oki Electric, Seiko, and other Japanese electronics companies are competing fiercely to develop new miniature products, such as pocket computers, ultracompact air conditioners, and hand-held copiers. They are pushing the physical limits of million-bit-per-chip (megabit) memory devices, which are being designed into pocket-size electronic memos and wristwatch televisions. As noted by Korean writer O-Young Lee, author of *Smaller Is Better: Japan's Mastery of the Miniature:* "In Japanese, the generic word for craftsmanship, *saiku,* translates literally as 'delicate workmanship.' In other words, to craft something is to make it smaller and fashion it delicately."

Miniaturization is only one Japanese tradition with industrial applications. A review of traditional Japanese folk arts reveals many other aesthetic principles with business applications, as shown in Table 4–1.

Bonsai, for example, is a spiritual and aesthetic art that teaches the basic principles of economy and controlled growth. Raising *bonsai* plants requires acute visualization, planning, and patience—skills that can be applied to bioengineering, for example, in which cells must be carefully cultivated, pruned, and grown into previously envisioned patterns. Three-dimensional computer-aided design (CAD) systems are being used for genetic modeling on both sides of the Pacific Ocean. In the field of biosensors, for example, proteins or other molecules are deposited layer by layer over silicon substrates to create multidimensional biostructures. In bioceramics, world-leading ceramics maker Kyocera is growing biological materials (called Bioceram) onto ceramic bones to develop artificial limbs that can be reattached to normal bones. These experiments are biochemical versions of *bonsai* growing.

Japanese gardens involve the creation of miniature, naturalistic spaces that can be useful for designing contemplative research areas. Aesthetically pleasing research parks and science

TABLE 4–1
The Cultural Foundations of Japanese Creativity

Traditional Art	Aesthetic Principle	Business Application
Wood carving (*netsuke*)	Miniaturization Animism	Pocket televisions Video animation
Bonsai	Miniaturization Trained growth	Electronic products Bioengineering
Flower arrangement (*ikebana*)	Creative forms Naturalism Asymmetry	Robot design Commercial landscaping Amorphous crystal growth
Rock gardens	Reductionism Aesthetic asymmetry Meditative space	Home construction Science city design Research lab design
Architecture	Multi-purpose rooms Open to nature Natural materials	Apartment housing Office complexes Office interiors
Paper folding (*origami*)	Manual dexterity Complex 3-D forms	"Transformer" toys Computer-aided design
Hand-sewn juggling balls (*otedama*)	Aesthetic play	Educational toys
Abacus (*sorobon*)	Manual dexterity Visualization	Calculator keyboards Computer simulation
Chopsticks	Manual dexterity	Robot fingers
Folding fans (*sensu*)	Collapsible space Aesthetic function	Laptop computer design Ergonomic furniture
Japanese characters (*kanji*)	Visualization Image recognition	Fifth-generation computer Visual scanners
Wrapping cloth (*furoshiki*)	Multi-purpose; compact	Folding solar panels

cities are already being built throughout Japan, such as the Kansai Cultural Research City near Kyoto and the Kibi Highlands Technopolis in the hills of Okayama City. Within research complexes, Japanese gardens offer meditative spaces for researchers in the Tsukuba Science City and surrounding research parks. And the naturalistic, open-to-nature features of traditional Japanese architecture are being used to create the illusion of spaciousness and calm in crowded offices and housing complexes. Mazda's new research center in Yokohama features a spacious meditative pool in the center of an open courtyard, reminiscent of the Zen sand garden at Ryoanji in Kyoto.

From childhood, Japanese are taught a variety of folk crafts and games that inculcate aesthetic and neuromuscular skills that are useful in business and research. Paper folding (*origami*), for example, not only requires manual dexterity and patience, but also visual acuity. Making the traditional "thousand cranes," which represents a long life, teaches Japanese children that with some imagination and hard work, even simple materials can be transformed into beautiful, complex three-dimensional forms. When folding paper, one must visualize the object to be created, a skill that is transferable to computer-aided design and simulation modeling.

The abacus, adopted from China in the sixteenth century, is another useful tradition. Calculating by abacus is taught in Japanese elementary schools and vocational high school accounting and business courses. Today, the abacus is still used in Japanese companies and banks because it is faster than calculators for addition and subtraction. Highly skilled abacus users can even calculate in their heads without using an abacus! These visualization and computational skills have potential applications in computer programming and simulation. Car designers, for example, use their computational skills when running computer simulations. These skills are useful in a wide variety of design and software industries.

Like origami, folding fans (*sensu*) have instilled the notion of collapsible space, which is useful for designing products such as laptop computers, modular furniture, prefabricated bathrooms, and multi-functional kitchen shelves. Because of high land costs and crowded conditions, Japanese furniture and tools are usually smaller and more collapsible than their U.S. equivalents. Mitsubishi Petrochemical, for example, introduced in 1986 paper-thin TEMAKI speakers that can be hung flat on the wall, rolled into cone shapes, printed on, or rolled up. The design of the speakers is based on the *temaki sushi* (*sushi* rolled in sheets of dried seaweed). The manufacturer Onmarugo, a maker of illuminated advertising signs, was seeking new interior decor products.

Finally, the Japanese language offers rich possibilities for new product ideas. Called "the devil's tongue" by early European visitors, the language can be extremely frustrating to Westerners because of its intricate, complex *kanji* characters. But like Chinese, learning Japanese trains one in complex pattern recognition, visualization, and recall—skills that are valuable in computer-aided design. Moreover, the complexity of Japanese *kanji* has forced computer makers to develop complex word processing software, facsimile machines, optical scanning devices, computer translation systems, and other advanced data-processing technologies. Although it has slowed Japanese computer development, the skills and technologies acquired along the way are valuable for future artificial intelligence (AI) and optocomputing systems.

We have explored the cultural origins of Japanese creativity and developed a new theory of creativity—the mandala of creativity. Now let's examine the management techniques that Japanese companies are applying to new product development at each phase in the mandala.

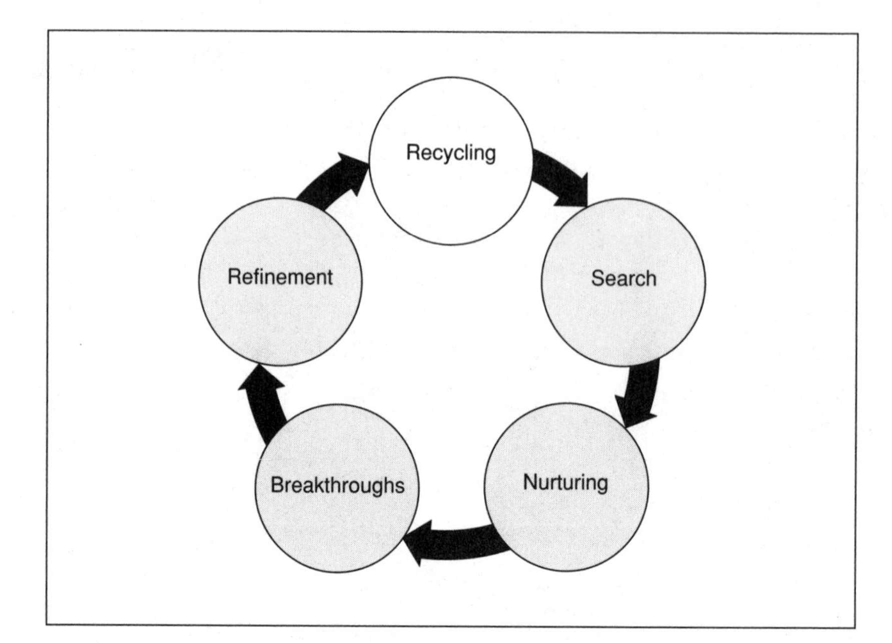
Recycling
Search
Nurturing
Breakthroughs
Refinement

再利用
SAIRIYO

5

Sairiyo: Recycling the Past

> He saw his escape in the Shino jar. He knelt before it and looked at it appraisingly, as one looks at tea vessels.
>
> —Yasunari Kawabata
> *Snow Country*

IN Kagoshima Prefecture on the southern island of Kyushu, there is a fascinating project being conducted at the Fine Ceramics Product R&D Institute, which is located at the New Artcraft Village in the Kokubu-Hayato Technopolis. Kyoto Ceramics (Kyocera) is developing new industrial concretes and building materials, ceramic electronic components, durable earthenware, ceramic engine parts, and artificial bones and teeth by combining its high-tech ceramic processing and business know-how with the centuries-old knowledge of materials and molding and firing techniques developed by the traditional Satsuma pottery makers. Their joint exploration is a marriage of old and new technologies.

But more important than the high-tech ceramics being developed by this consortium are its implications for the future. The Kagoshima experiment, which is rediscovering forms of creativity rooted in Japanese traditions, symbolizes Japan's attempt to revive its declining industries with an infusion of new ideas and technologies. Drawing upon their cultural origins, Kyocera and its partners are forging a new materials revolution that will reshape the industrial world in the twenty-first century.

The Japanese are masters of recycling old technologies. For

products as diverse as soy sauce, earthenware, and textile dyes, they are constantly seeking ways to create new products and industries out of the garbage heap of history and dying industries. While Americans throw away old ideas, Japanese prize them like fine wines. Good ideas, like land and gold, are scarce commodities.

It is often said that Europeans look to the past, Americans discard their past, and Japanese carry their past with them into the future. From the time they are born, Japanese are constantly reminded by their family, friends, teachers, and employers to conserve their limited resources. "*Mottainai*"—what a waste! These are powerful words, with strong emotional undercurrents because of their association with Japan's past famines and its suffering after World War II. Living on a small island with few natural resources, Japanese children traditionally were taught never to waste anything, not even scraps of paper or pieces of string. They were encouraged to recycle and reuse everything and to find new uses for seemingly useless objects. The practice of trading old newspapers and cardboard for toilet paper—or *chirigami kokan*—is familiar to most Japanese. Even today, one often sees old men pulling rundown paper collection wagons through downtown Tokyo and Osaka. Although the habit of recycling is waning in an era of growing prosperity, it is still deeply embedded in the Japanese psyche. Sudden, unexpected events, such as the 1973 oil crisis, the 1985 yen shock, the 1987 stock market crash, and the growing environmental crisis remind the Japanese of their economic vulnerability to external forces. They must save and recycle in order to prosper and prepare for future shocks.

This recycling ethos is not as commonplace in the West, and especially not in the United States, which is endowed with abundant natural and material resources. Whereas American companies expend large amounts of financial and human resources pursuing major breakthroughs, Japanese companies have traditionally sought the security of recycling proven ideas and technologies. It is an inherently conservative approach, but very effective. What are some of the methods used in Japan today? How are they implemented by high-tech companies?

Hybrid Technologies

Combining old ideas with new ideas, or combining two old ideas, is a favorite Japanese method for recycling old products and technologies. When faced with stagnant markets, Japanese companies often review their patents, products, and technologies to identify new product ideas. Unlike U.S. companies, which prefer to hire experts to develop specialized products, Japanese companies seek generalists who are rotated through departments to encourage cross-training and the free flow of ideas. For example, when Sony began losing market to South Korean videotape recorder (VTR) makers in 1986, researchers from different divisions sat around thinking of ways to use Sony's VTR technology in new video products. After some market testing, Sony introduced a new hybrid product—a large-screen TV console with built-in VTRs and stereo speakers.

Since World War II, hybrid technologies have been a popular way to recycle old technologies. Japanese textile makers, for example, combined their dye techniques with advanced chemical processes to diversify into industrial dyes. Matsushita, Sony, Toshiba, and other video makers are combining their television, VTR, stereo, CD, and telephone technologies to create videophones, stereo televisions, and high-definition television (see Figure 5–1).

For the office market, Canon and Ricoh are combining facsimile machines with telephones, copiers, and printers. One very creative idea is the electronic copyboard—a combination of blackboards and facsimile machines that was developed by Oki Electric and Plus Company. Images written on the surface of the copyboard are transmitted through a lens to a visual memory chip, which then issues a photocopy. In computing, Sony is using its color TV know-how in high-resolution graphics terminals and engineering work stations. Its Walkman was the idea behind the "Videodiscman"—a 3.5-inch optical disk player for use as computer storage. For Japanese companies, the potential for creating new hybrid technologies is endless, limited only by their imagination.

FIGURE 5-1
The Japanese Approach to Hybrid Technologies

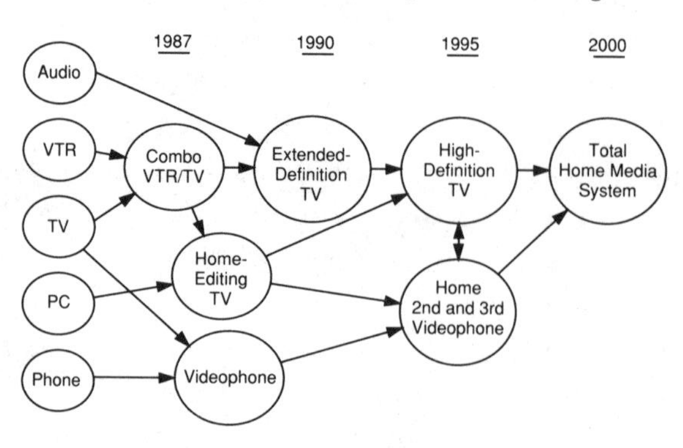

SOURCE: Dataquest, Inc.

Fusion Technologies

Whereas hybrid technologies combine two ideas without changing their essential nature, fusion technologies effect a basic transformation of the technologies involved. *Yugo-ka*—the fusing of various ideas and technologies—is a popular approach in Japan because it leverages the Japanese propensity for long-term planning and teamwork. The method involves bringing together multidisciplinary teams of researchers to explore systematically new combinations of existing technologies.

Fumio Kodama, director of science policy at the Science and Technology Agency (STA), observes that Japan's mechatronics industry (the merging of machine tools and electronics, such as computerized numerical control manipulators) was an early example of technology fusion.

> The mechatronics revolution in Japanese machine tools first became possible through cooperation between Fanuc—a spin-off of a communications equipment manufacturer—which developed the controller, NSK—a bearing manufacturer—which developed the ball screw, and the materials manufacturer, which developed the Teflon material used to coat the sliding bed.

In 1971 the Japanese government passed the Law for Provisional Measures to Promote Electronic and Machinery Industries (*Kidenho*), which encouraged "the consolidation of machinery and electronics into one," thus giving rise to the mechatronics industry.

In 1986 MITI's Industrial Structure Deliberation Council issued a white paper, "The Fundamental Outline of Twenty-first Century Industrial Society," in which it introduced the concept of technology fusion. MITI proposed a new method—the "interindustry technology fusion index"—that identifies the level of research spending by various industries in certain product, or "technology fusion," areas. As shown in Figure 5-2, areas with a strong overlap indicate potential areas of research synergy. For example, MITI believes that excimer lasers and ion implantation technology can be used in the manufacturing of propellers and X-ray mirrors. The goal is to help chemical, metals, machinery, electrical, and other heavy industries to diversify into new fusion technologies. MITI's philosophy is

FIGURE 5-2
Interindustry Technology Fusion Index

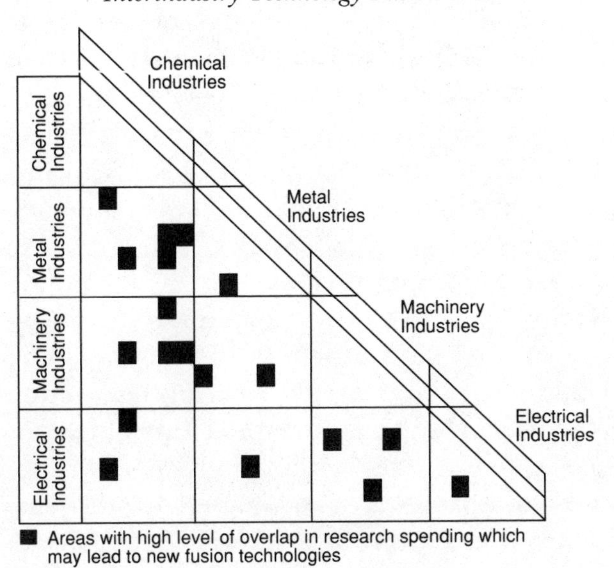

SOURCE: Industrial Structure Deliberation Council, "The Fundamental Outline of Twenty-first Century Industrial Society" (26 May 1986).

reflected in the subtitle to its white paper: "Towards the International Harmonization of the Industrial Structure and Fusion of Creative Knowledge." In 1988 the Japanese Diet sanctioned MITI's technology fusion policies by passing the Fusion Law, which empowered MITI to establish technology exchange plazas, fusion management research centers, regional fusion centers, and "catalyzer" industry research.

MITI's fusion technology policies are being initially implemented in biotechnology; distillers such as Asahi Brewery, Kirin, and Suntory are combining their fermentation know-how with new biotechnology techniques to develop new food preservatives, liquors, and medicines. In 1985 MITI's Biocomputer Project drew experts together from fields as diverse as biotechnology, software, chemistry, computer architecture, and chip design. The goal of this project is to develop next-generation biosensors and bioelectronic systems. Sanyo Electric, one of the participating companies, is using biotechnology, chemistry, software and AI, chip design, computer system architecture, and CAD tools to explore the new field of bioelectronics. This technology fusion approach is shown in Figure 5–3.

Even for the group-oriented Japanese, managing technology fusion is a difficult, time-consuming process. Multidisciplinary teams with the flexibility and willingness to explore totally unrelated fields are needed; overspecialization is a serious handicap. According to Mitsuo Mizukami, R&D manager of Sanyo's Tsukuba Research Center: "It took a year just for our bioelectronics researchers to understand each other's fields. But now we have discovered common ground and are working on pattern recognition, optical chips, and biocomputing." Eventually, Japan's top electronics companies hope to develop biocomputers that mimic the neural and sensing functions of human beings. They are already merging bioelectronics with the new fields of neural networks and fuzzy logic to develop these biocomputers.

Japanese companies are pursuing many other fusion technologies, including optocube computing (massively parallel optical computers using Trinitron-like switching guns to transmit laser beams from one grid of fiber-optic cables to another),

FIGURE 5–3
Sanyo's Technology Fusion Approach to Bioelectronics

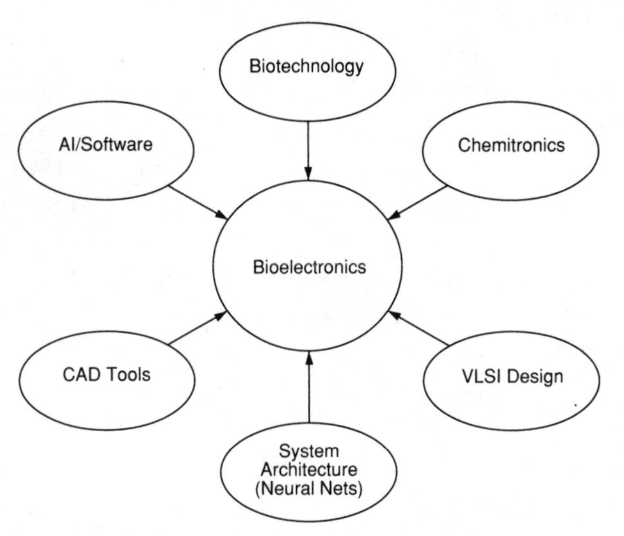

SOURCE: Dataquest, Inc.

optomechatronics (the marriage of optical, mechanical, and electronic engineering), bioceramics (biotechnology and ceramics), and biocommunications (biotechnology and communications networking).

In the 1990s we may see major Japanese breakthroughs in fusion technologies. As STA's Fumio Kodama notes: "A single technical breakthrough alone is not sufficient for progress in high technology. Rather, only through the organic fusing of several technical breakthroughs in a number of different fields can a new technology be created."

Crossover Technologies

Another popular Japanese method for creating new products is to transfer an idea or technology from one industrial sector to another. Because Japanese workers tend to be generalists who move freely back and forth between divisions, they are able to transmit ideas rapidly between consumer and industrial product areas. Companies institutionalize crossover research by rotating their employees every two or three years through different divisions and building multidisciplinary product development teams.

Sony developed light-sensitive sensors known as charge-coupled devices (CCD) for its 8mm home video camera, then put CCD sensors into office security cameras, videophones, and document scanners for optical storage equipment. Matsushita and other audio equipment makers transplanted their CD players from the home to the car. And Japanese robot makers are diversifying into fire-fighting robots, hospital nurse robots, underwater robots, and even sushi-making robots. In each case, applying a crossover technology required the ability to cross technical and corporate lines.

Extracting the Essence

Frequently, Japanese companies take the essence of a product idea to develop a totally new product. After the 1985 yen shock sent Japanese exports plummeting, Matsushita gathered a group of female engineers, product designers, market researchers, and interviewers to explore the possibility of new home electronics markets. After conducting extensive interviews with working women and housewives, the group discovered market demand for an appliance that could bake bread overnight.

In the postwar period, many Japanese have shifted to toast and coffee for breakfast because it is lighter and more convenient than traditional miso (soy bean) soup, pickles, and rice. In addition, many Japanese returning from vacations and business trips abroad have developed a taste for freshly baked breads. If Japanese cooked their own rice, why could they not bake their own bread? Incorporating ideas from customer interviews, the Matsushita team designed home bread-baking machines in which dough is placed in the evening, automatically baked overnight, and ready as fresh bread in the morning.

In other words, the researchers at Funai Electric and Matsushita identified the essence behind the popularity of electric rice cookers—the convenience and speed of preparing food in a small home appliance—and then applied the same idea to the development of an automatic bread-baking appliance. A stunning success in the Japanese market, Matsushita's bread-baking machine was a hit at the 1988 Consumer Electronics Show in Chicago and is now being exported.

Another example of extracting the essence is the "capsule hotel"—small prefab sleeping compartments stacked one above the other near busy train and subway stations. The idea originated from the Pullman sleeping compartments on trains. Because of late working hours and evening entertaining, many Japanese businessmen miss the last train home or are too tired or drunk to get home unassisted. To save time, they rent small sleeping capsules at 20–50 percent less than regular business hotel rooms. These capsules are easy to clean and maintain and are outfitted with a color TV, radio, alarm clock, reading light, and small shelf. For harried businessmen on the go, they are convenient quick-stop hotels.

Process Spin-offs

One sure way of recycling ideas is to spin off new products from existing industrial processes. Japanese companies, whether they make chips or medicines, take pride in their manufacturing processes, which they view as a fountain of new product innovations. By developing and improving their processes, they can ensure a steady stream of creative new products.

Memory chip technology is a good example. Since capturing over 70 percent of the world market in the early 1980s, Japanese companies have pushed the limits of memory process technologies and are now developing specialized memory chips for high-definition TVs (HDTVs), telephone message playback, laptop computer memory storage, and other new applications. They are modifying their process technologies, but more important, offering a variety of designs to take advantage of their process strengths.

HDTV, which will double the clarity of TV images, is an emerging technology that will create numerous spin-off product opportunities, as shown in Figure 5–4. In the mid-1970s, color television made possible Sony's Betamax and Matsushita's VHS videotape recorders and movie cassettes, which have blossomed into multibillion-dollar spin-off markets. During the 1990s, HDTV will spin off totally new markets, such as TV program editing, stereo film cassettes, and videotheaters. Like memory chips, these spin-off products require massive investments in process innovations. But once made, these process innovations yield new products and markets, such as video retailing and film production.

Sandwiching Ideas

Sandwiching new ideas between several old ideas—or vice versa—is another way Japanese companies recycle ideas. Sony's Walkman is perhaps the classic example of this approach. In *Made in Japan,* Akio Morita explains how his partner Masaru Ibuka came up with the idea of putting standard-size headphones on Sony's portable stereo tape recorder. Between these

FIGURE 5–4
Television Spin-off Markets

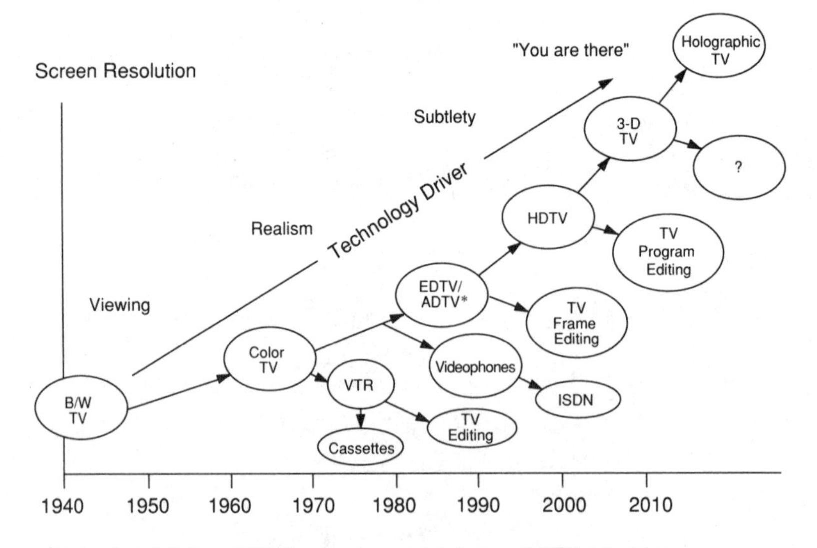

*Extended-definition (EDTV) and advanced-definition (ADTV) television.

SOURCE: Dataquest, Inc.

two old ideas, Sony sandwiched a new idea—personal stereo sound. Morita sensed that a market existed for the Walkman from personal observation: "In New York, even in Tokyo, I had seen people with big tape players and radios perched on their shoulders blaring out music."

Another example is MITI's three-dimensional chip project in which it is developing "high-rise" chips. Existing chips are two-dimensional, that is, their circuitry is drawn only on one plane. In the future, the Japanese plan to develop chips that feature memory, logic, and laser functions on different levels. To make his point, MITI research manager Hiroshi Kawakami used the analogy of a Japanese train station:

> Usually, common activities such as buying tickets are done on the first floor, boarding platforms are located on the second floor, and restaurants and shops on the third floor. Likewise, our 3-D chips will put only common functions on the main level and less-used functions on other floors. Wiring will be found overhead—like in the ceiling of a room.

By seeing the analogy between three-dimensional chips and railroad stations, MITI researchers have sandwiched multiple functions—memory, logic, and lasers—onto one high-rise chip.

Thus, the Japanese use many techniques to recycle existing ideas. But when they run out of existing ideas, how do they explore new ideas? Let's look at the idea search techniques being developed in Japan today.

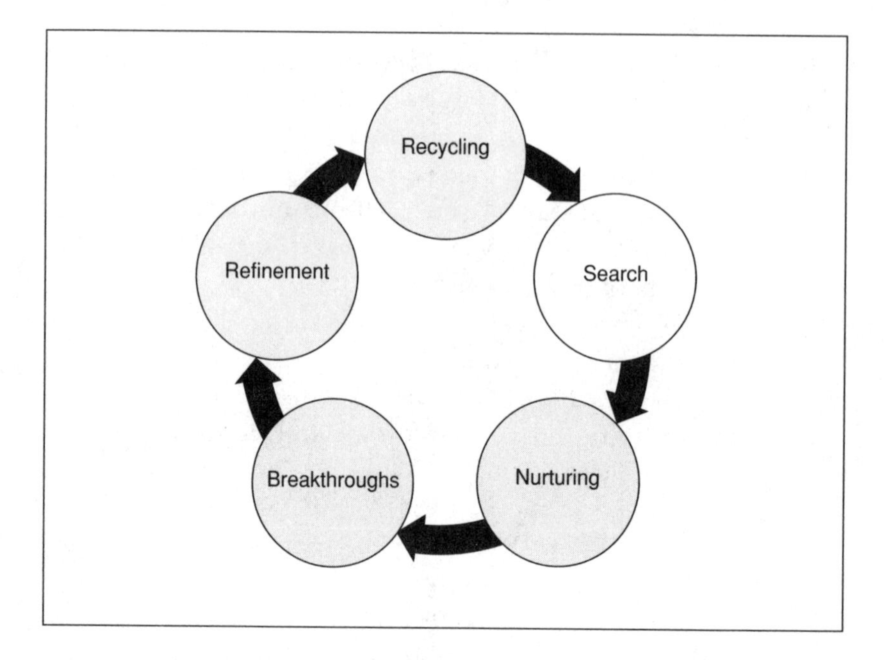

TANSAKU

6

Tansaku: Exploring New Ideas

Curiosity is the key to creativity.

—Akio Morita
Chairman, Sony Corporation

As Japan catches up with the West, its scientific "free ride" is coming to an end. Many nations now insist that Japan expand its own exploratory research. The Japanese realize they must contribute to the pool of scientific knowledge, not just take and commercialize foreign technologies. Makoto Kikuchi, director of the Sony Research Center, is adamant about Japan's need to pursue creative research: "The level of Japanese technology has risen, and from now on Japan, as a leading technological state, can no longer continue to simply take foreign technology and advance it a few steps further down the technological track. It is now time to conduct independent research that is uniquely Japan's own."

Old habits die hard, and exploring revolutionary new ideas will not be an easy task for the Japanese. Creative research requires a great deal of curiosity, adventurousness, and risk-taking. While Westerners often relish the personal challenge of frontier research, most Japanese, who prefer the comfort of proven ideas, shy away from exploring risky technologies on their own. They prefer a more structured, team-oriented approach to avoid losing face or being left out.

Yet Japanese companies are beginning to encourage exploratory research. What methods are they using? And how are they being implemented? This chapter examines six popular methods used in Japan today: visionary thinking, global search, spiral development, matrix frameworks, technology trees, and technology road maps.

Visionary Thinking

In 1969, when Apollo 11 landed on the moon, Japanese leaders were awed by the impressive brainpower and imagination that Americans could unleash by pooling their efforts in pursuit of a common vision. When President Kennedy vowed "to put a man on the moon by the end of the decade," they were skeptical. But when the mission was accomplished, they realized the power of visionary thinking.

Since the early 1970s, the Japanese government has issued its own "visions of the future" to encourage private industry to become more creative and visionary. In November 1974 MITI called for a knowledge-intensive industrial structure. Not as compelling or dramatic as the Apollo Project, it nevertheless gave industry leaders a down-to-earth vision to rally around and a direction toward which they could steer their research efforts. In 1980 MITI issued its famous "Visions for the 1980s," which called for a shift to more creative research and targeted fourteen high-technology industries for accelerated development: aircraft, space, optoelectronics, biotechnology, computers, robotics, medical electronics, semiconductors, word processors, new alloys, fine ceramics, medicine, software, and mechatronics. This vision statement provided the impetus for Japan's national research projects, including the fifth-generation computer, supercomputer, FSX fighter plane, and optoelectronics projects.

MITI's "Visions for the 1990s" will soon be issued and is likely to address the pressing issues facing Japan today: economic globalization, domestic demand expansion, the rise of trading blocs, basic research, land reform, educational reform, changing values and lifestyles, the aging society, and Japan's growing global responsibilities.

Contrary to popular belief, MITI's vision statements are not just the brainstorm of MITI bureaucrats, but the consensus of many opinion-makers who sit on MITI's Industrial Structure Council and numerous study committees. The visions are aimed at a broad audience: depressed industries diversifying into high technology, small companies struggling with the strong yen, high-growth companies seeking new markets, consumers, labor unions, environmentalists, and home-owners. As in the past, MITI uses a variety of policy instruments to implement these visions, including R&D subsidies, tax incentives, special depreciation allowances, Japan Development Bank (JDB) loans, research program funding, export credits, import tariffs to protect infant industries, antitrust exemptions, government procurement, and "administrative guidance," or behind-the-door negotiating.

MITI's Fifth Generation Computer Project is an example of how the Japanese government uses its policies to influence the direction of scientific research. Prior to the project, Japanese computer companies spent relatively little money on AI, parallel processing, and relational data bases, all of which were pioneered and developed in the West. Yet, after MITI funded the project, Japanese companies began increasing their investments in these fields. Fujitsu and Hitachi, for example, developed expert systems to design very large-scale integrated (VLSI) chips, while Matsushita and Mitsubishi introduced AI-based flexible manufacturing systems. Japanese hospitals, whose systems, by Western standards, were seriously outdated, began experimenting with medical expert systems. The Fifth Generation Computer Project triggered advanced research among ordinarily risk-averse Japanese companies.

This use of policy visions as a catalyst is often misunderstood in the West. Many Westerners, for example, believe the primary purpose of MITI's Fifth Generation Computer Project was to overtake the West. But the project appears to be aimed more at breaking the "tyranny" of alphanumeric data entry, which is based on Western languages, and developing new image-oriented computers more suitable for the Japanese language. J. Marshall Unger, professor of Asian Languages at the University of Hawaii and author of *The Fifth Generation Fallacy,* observes that the Japanese written language is so complex that it has delayed software development and burdened the Japanese with extra costs. By pumping money into AI research, the Japanese hope to overcome their software shortage and raise white-collar productivity. In this sense, the project is MITI's experiment in "catch-up" creativity. The Fifth Generation Computer Project may fail to "reach the moon," but it is already spinning off new products and ideas that have immediate commercial value, such as expert systems, parallel processing, and Japanese language software.

The visions approach, however, has its limitations. When MITI exerted greater financial control over Japanese industry, its industrial policies could be implemented through project funding. Major Japanese corporations now have large financial assets and can easily raise funds in international capital markets. Moreover, they now rely more heavily on their own research laboratories and joint R&D projects with foreign partners to stimulate their creative juices. MITI's vision statements will play a lesser role as Japanese companies become more independent and issue their own vision statements.

Global Search

Although traditionally weak in scientific exploration, Japanese corporations are masters at searching globally for new ideas and technologies. They search for answers even when they do not fully understand the problems they hope to solve (see Figure 6–1). In *Japan as Number One,* Professor Ezra Vogel describes the Japanese global search process:

FIGURE 6–1
Comparison of U.S.–Japanese
Corporate Technology Search Approaches

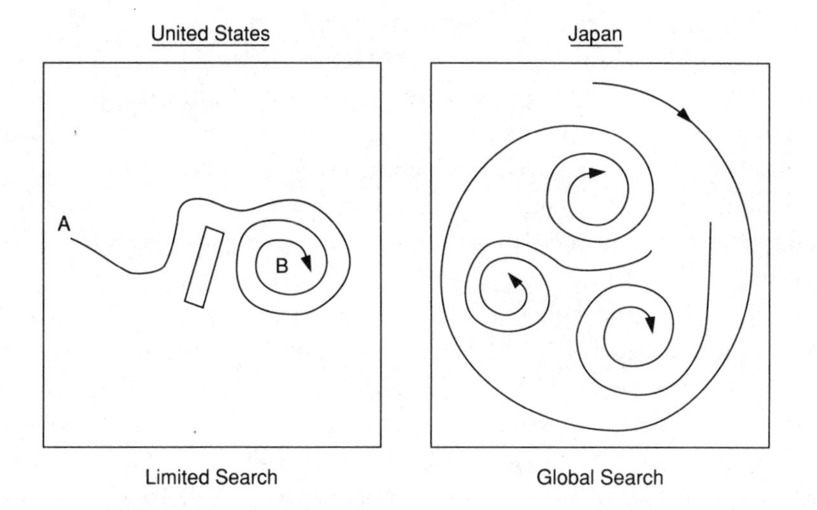

> Organizations send out observation teams and invite in experts. They gather information from classrooms and golf courses, from conferences and bars, from think tanks and television. They gather it from professionals and amateurs, friends and foes. New friends are cultivated because they might provide access to information. Potential sources are carefully nurtured so that further queries can be processed as needed. New areas of knowledge are *explored* [emphasis added] to provide new clues, and people are assigned to spend several years mastering potentially profitable specialties. The process is nothing if not thorough.

As Westerners have learned, Japan's global search method is a powerful business tool. By exploring unfamiliar territory and leaving few stones unturned, Japanese companies uncover many new ideas and technologies with practical applications. Moreover, by probing assiduously and asking difficult questions, they often arrive at creative solutions to complex problems. As a popular adage goes: Asking the right questions is half the battle.

Sharp's solar-powered calculator exemplifies the creative potential of the global search method. In the late 1970s, Sharp identified a market need for battery-less calculators and began searching for inexpensive, efficient solar energy technologies that could be used to create a replacement for bulky photovoltaic cells. After investigating universities and companies working on solar energy generation techniques, Sharp identified a small company in Troy, Michigan, named Energy Conversion Devices (ECD) that had developed a novel amorphous silicon material that could be mass-produced in continuous sheets like rolled steel. Unlike existing silicon materials being used at the time, ECD's amorphous silicon materials featured irregular crystal lattices with highly efficient energy conversion properties. The new technology was exactly what Sharp needed for its solar-powered calculator panels. Although ECD was overlooked by major American calculator makers, Sharp's Tadashi Sasaki took a gamble and used its amorphous solar cell in Sharp calculators. The gamble paid off: Sharp—along with Casio—became a major player in the ruthlessly competitive calculator business, which is now dominated by solar-powered calculators from Japan.

Although U.S. companies had been exploring battery-less solutions, they not only suffered from the NIH syndrome but ignored ECD because of the unorthodox background and ideas of its president, Stanley Ovshinsky. He is a high school graduate dismissed by many companies as a maverick who is unable to make his company profitable. Sharp was not blinded by its pride, even though it is Japan's leading optoelectronics device maker, and it has profited handsomely from ECD's invention. As Stanley Ovshinsky observed in his television biography, "Japan's American Genius": "The Japanese are very good at accepting new ideas, and acceptance of a new idea is an essential part of creativity."

Spiral Development

While global searches are useful for seeking ideas and technologies outside the organization, many Japanese companies create new products internally through a cyclical process. The spiral development wheel, which was borrowed from the West, is becoming a popular creativity technique. Product development is not seen as a linear process from basic research to final distribution, but as a recursive feedback process that involves the interaction of four major divisions in a company: basic research, product development, manufacturing, and sales and distribution.

Toshiba's best-selling laptop computer is an example of the spiral development approach. The idea began in the late 1960s, when Toshiba identified the need for a Japanese-language typewriter capable of handling 3,000 Japanese characters. Toshiba initially developed a tablet-input typewriter in 1968, but the product failed in the marketplace. To develop better language capabilities, Toshiba assigned a young researcher to attend a Japanese university to study basic linguistics with a noted professor and funded research in *kana-kanji* conversion (turning phonetic letters into complex characters) informally for ten years. In 1978 the development of dot matrix printer technology made Japanese word processors feasible. Toshiba incorporated these technologies into its *kana-kanji* Japanese word processor. But office distributors asked for smaller, lighter word processors, so the machine was sent back to the laboratory for more research. While Toshiba was developing new chip technology to make smaller word processors, new technologies emerged. Thermal printers and liquid crystal displayers (LCDs) suddenly made Japanese word processors portable and attractive.

In the mid-1980s Toshiba introduced its laptop word processor, Rupo-1, which took off in the Japanese marketplace. Despite U.S. sanctions on Toshiba imports after its shipment of propeller grinders to the Soviet Union, Toshiba cracked the tough U.S. computer market by licensing U.S. software and producing laptops in southern California. But the story does not end there. Toshiba is now researching voice recognition and synthesis to develop voice-activated word processors in the 1990s.

Why does the spiral development method work so well in

Japanese companies such as Toshiba? There are several reasons. One is the longer term investment horizon of major Japanese companies, which are willing to wait many years before achieving a healthy return on investment (ROI). A second, more compelling, reason is the relative flexibility of the Japanese labor force. In most Japanese corporations, rigid job definitions are nonexistent and employees are encouraged to work outside of their areas of specialization. People are freely rotated from job to job. Moreover, product champions are allowed to fail without the fear of being laid off. But most important, Japanese view products and services as incomplete (*mikansei*), requiring constant improvements and refinements (*kaizen*). These are not uniquely Japanese principles; they can be learned by any company with a willingness to experiment and learn.

Matrix Frameworks

Global searches and spiral product development are useful for incorporating existing ideas into new products, but they are inadequate for exploring new frontiers. Increasingly, Japanese companies use matrix frameworks—technical or management charts that plot the status of scientific developments and technology trends worldwide—to structure and identify promising areas of research, or "research holes." These frameworks provide "handlebars" for shifting from the known to the unknown and for helping Japanese organizations decide where to deploy their research funds for maximum return.

Superconductivity provides an example of the strategic power of matrix frameworks. The temperature at which materials become superconducting (critical temperature) has gradually risen since the discovery of superconductivity in 1911, opening the door for commercial use. In late 1986 researchers at IBM Zurich discovered high-temperature superconducting materials that won them the Nobel Prize. The discovery triggered a worldwide search for new superconducting materials, but after a rapid rise in critical temperature, new discoveries began to level off.

Tokyo University researchers plotted the discoveries on a chart and noticed a new trend emerging: faster progress was being made with certain materials, such as barium, bismuth, and strontium oxides, than with more popular materials, such as yttrium and lanthanum. Following their instincts, researchers at the National Research Institute of Metals (NRIM) in the Tsukuba Science City outside of Tokyo discovered a new ceramic oxide that became superconducting at −243 degrees Celsius, over 50 degrees higher than previous records. This record was later surpassed by Professor Paul Chu of Houston University. Nevertheless, the Japanese discovery stimulated research into totally new superconducting compounds. As superconductor expert Robert J. Cava of AT&T Bell Laboratories advised superconductor researchers after the Japanese discovery: "Watch the trend lines." When properly used, matrix frameworks can uncover trends often overlooked in the heat of competitive research.

Technology Trees

The Japanese have traditionally been great lovers of natural beauty, as expressed in their rock gardens and *bonsai.* Symbols of this love of nature carry over to their search for new ideas and technologies. One method used by high-tech companies is the technology "tree": a basic idea, or "seed" technology, evolves into many application "branches" that are cultivated to develop into entirely new industries. These branches are grown by exploring new applications that are conceived by in-house researchers or announced by competing companies. Over time, the technology tree develops from a simple tree with a few application branches into a complex tree with dozens of branches.

Sharp, a major audio electronics maker, uses technology trees to explore new audio technologies. As shown in Figure

6–2, the digital audio tree is rooted in a variety of seed technologies, such as digital recording, VLSI chips, and optical technologies. These basic technologies branch out to a variety of new audio products that are "pruned" by Sharp research and marketing divisions. Pulse-code modulation (PCM) technology, for example, promises higher quality audio sound, leading to the possibility of PCM cassettes, recording machines, and adapters.

By using technology trees, Sharp is able to take existing ideas and systematically explore their possible new applications. Technology trees are useful because they vividly depict the organic development of new technologies.

Technology Road Maps

While technology trees are useful for exploring the potential of new ideas and technologies, they have limited usefulness for portraying very complex technologies and ideas. Moreover, technology trees offer no time horizon for each technology. As a result, Japanese companies switch to technology road maps when it is possible to forecast technologies with greater predictability.

Technology road maps are much more systematic than the symbolic technology trees. Road maps show the timing and interaction of parallel technologies and provide a sense of direction and continuity for long-term investments. Japanese companies regularly use technology road maps to explore new ideas with commercial potential five to fifteen years in the future.

Figure 6–3 shows Sharp's technology road map for digital audio sound, speakers, communications, and digital audiotapes from 1982 to 1990. The road map indicates how Sharp envisions the development of CD players, still-image CDs, CD-ROM (read-only memory) disks, information disk files, and erasable-rewrite optical disks. By putting each technology into a specific time frame, Sharp managers can quickly muster the supporting

FIGURE 6–2
Sharp's Digital Audio Technology Tree

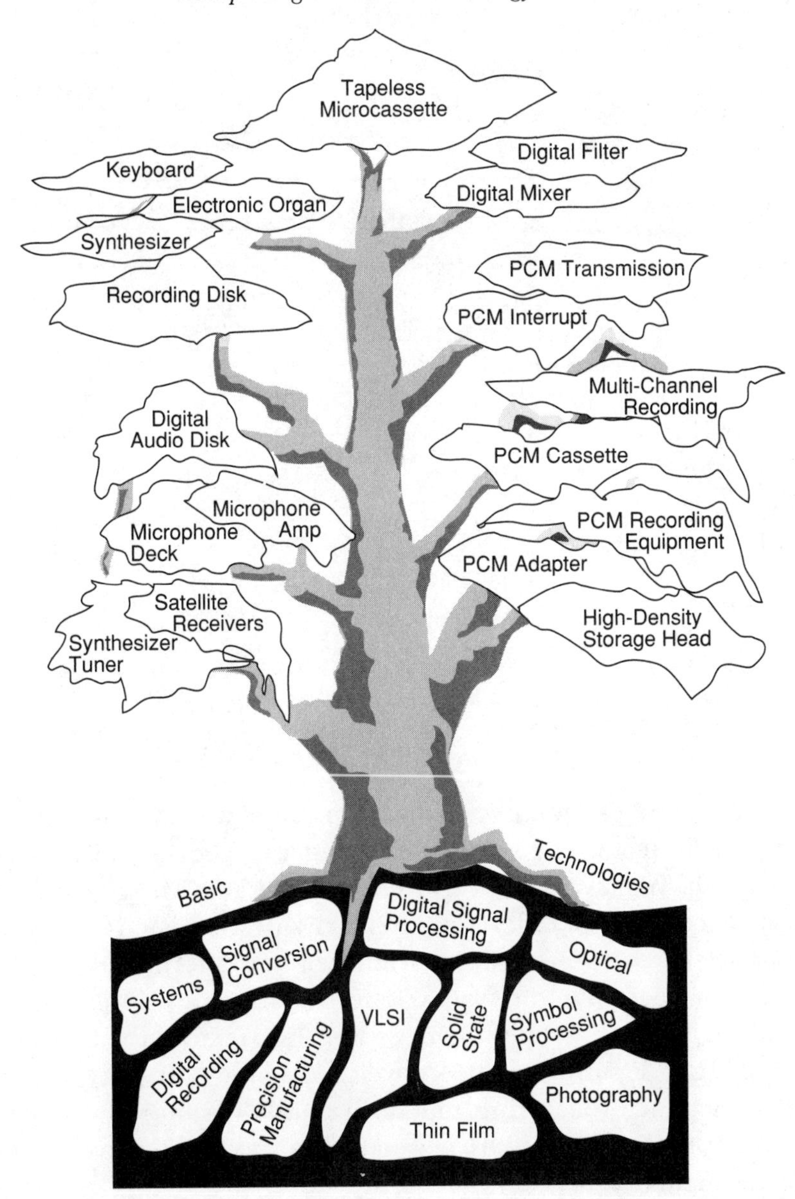

SOURCE: Sharp Corp. © Japan Management Association.

FIGURE 6–3
Sharp's Audio Technology Road Map

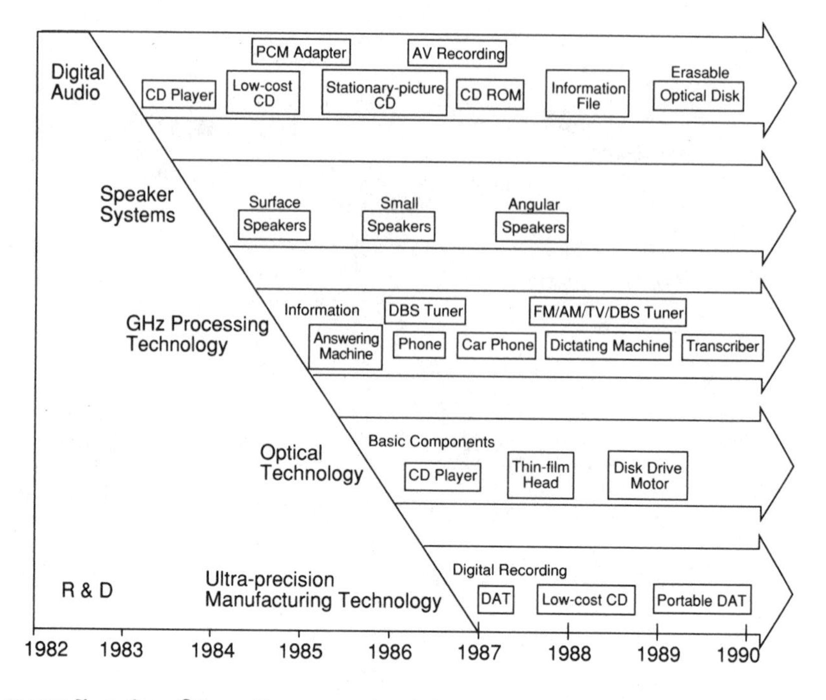

SOURCE: Sharp Corp. © Japan Management Association.

research and capital investment required to make the transition from one technology to another and can coordinate parallel research. When a technical breakthrough is made, researchers can accelerate parallel research activities; when a technical obstacle is encountered, they can seek new alternatives.

The road maps are valuable for identifying synergy among various products that might escape the attention of groups working independently. Moreover, road maps provide product planning, manufacturing, marketing, and distribution divisions a chance to anticipate changes in business direction and capital investments.

These are the major exploratory techniques being used by Japanese companies in their search for creative new ideas and products. Although these methods are also being used by Western companies, the Japanese are developing alternative approaches because of their intense interest in exploring new fields. Next, we will examine the "idea cultivation" methods being tested in Japan.

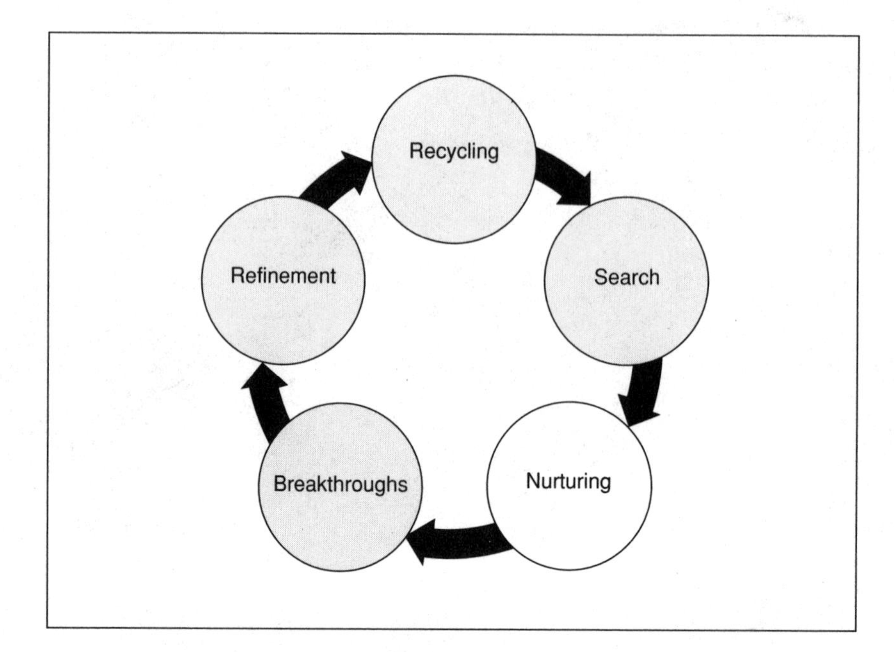

IKUSEI

7

Ikusei: Nurturing Creative Ideas

How can the next generation of managers give the *be* (youth) generation something meaningful in work, thereby revitalizing business?

—Noboru Makino
Chairman, Mitsubishi Research Institute

PSYCHOLOGISTS and scientific researchers are well aware that the germination period for creative ideas may last for several months, years, or, as in the case of high-temperature superconductors, even decades before a breakthrough occurs. In their pursuit of breakthrough ideas, Japanese companies are carefully planting and nurturing technology "seeds" with the intention of harvesting them in the twenty-first century. They realize that most breakthroughs are totally unintentional and unexpected, but that the groundwork can be laid by nurturing creative ideas. They have become the world's latest "idea cultivators."

Cultivating ideas is difficult in Japan, however, because of the tremendous pressures to bring products to the marketplace quickly. Commercializing ideas, not cultivating them, reaps the greatest rewards for companies in the hypercompetitive Japanese marketplace. As a result, pursuing unorthodox ideas is discouraged because it is risky and requires enormous patience and a willingness to swim against the tide. In Japan it is not "what is unique?" but "what are others doing?" that commands the lion's share of corporate funding and support. Only the most persevering people are able to withstand this intense pressure to conform and be "part of the team."

Nevertheless, Japanese managers and industry leaders are beginning to let their creative young people take risks and experiment on their own. These managers, aware that product imitation is the surest road to corporate decline, are the creative samurai who will help Japanese corporations make the leap to creativity.

Dr. Noboru Makino of Mitsubishi Research Institute is one of the foremost thinkers on creativity and innovation in Japan. In his book *Decline and Prosperity: Corporate Innovation in Japan,* he writes:

> The hierarchical structure [of the Japanese company] will have to be changed, independent venture businesses such as in-house and satellite businesses will have to be adopted, the seniority system will need to be abolished, and job mobility will have to be increased. . . . I feel more and more that the key to new product development lies in the personality of the team leader, not the organization, and thus the right person must be found to take charge of development. Such a person should also be able to work in total obscurity for a while, as product development is time-consuming and frustrating and results do not materialize overnight.

These are strong words of advice for Japanese managers, who for years basked in the awe and respect accorded them by Westerners. Yet they ring with an element of truth: if Japanese managers do not plant and cultivate technology seeds now, their companies will have little to harvest in the 1990s, especially as the enforcement of patent and intellectual property rights challenges imitators in the marketplace.

What are some of the ways that Japanese managers are cultivating new ideas? This chapter examines three common methods: after-hours socializing, designing creativity-inducing environments, and setting up venture business incubators.

After-Hours Socializing

Anyone who has spent time in Japan quickly discovers that the most creative ideas usually emerge after work in restaurants and bars. Walk into any bar in Tokyo, Nagoya, or Osaka in the evening and you will find groups of managers, engineers, salesmen, secretaries, and government officials clustered about telling jokes and avidly discussing everything from baseball and golf scores to the status of a company project. Besides cementing ties between coworkers and clients, after-hours socializing enables the Japanese to build important relationships and to get feedback from close friends and colleagues. In the bacchanalian heat of eating, drinking, and discussing, offbeat ideas are tossed about, expanded, picked apart, and playfully combined with other ideas. For many Japanese, evening is a playground for the mind and the soul—a welcome release from the daily grind of the office and factory.

To take advantage of this lively synergy, Japanese managers frequently take their research, marketing, and design teams to local bars and restaurants for a relaxing evening out. The real goal is to build group harmony, sweep away inhibitions, and get people thinking about new ways of doing things.

MITI's VLSI Project in the late 1970s is an example of how Japanese use after-hours socializing for maximum effect. In 1976 former MITI official Masato Nebashi was named executive director of the new VLSI Research Association, which was organized to develop high-density memory chips for computers and telecommunications equipment. After hiring 100 top scientists from rival companies, he encountered problems getting them to cooperate, so he organized regular meetings to discuss research results, rotated engineers through the project, and maintained close ties with the participating companies. But nothing worked. As William Ouchi describes it in his book *The M-Form Society:*

> Despite these efforts at team-building, the walls between companies and between lab sections remained

> thick. Nebashi finally resorted to "whiskey operations," taking small groups of scientists out for drinks in the evening, then defining and solving the problem areas that had been revealed the night before. After three years of work, the walls began to come down. . . . Conversations between sections were formally held each Saturday, and in the evenings, conversations would go on late into the night.

Nebashi's encouragement of after-hours socializing was instrumental in getting VLSI Project members, who had rarely sat down with people from rival companies, to communicate with each other more openly. The results were stunning. During the course of the project, member companies filed over 1,000 project-related patents in Japan. The Japanese share of technical papers at the prestigious International Solid State Circuits Conference leaped from 25 percent in 1981 to 44 percent in 1987. In 1982, only two years after the project ended, Japanese chip makers captured 70 percent of the world DRAM (dynamic random access memory) market. By 1987 major computer makers such as Apple, Tandy, and Compaq felt the pinch of Japan's domination of the DRAM market. In the critical one-megabit memory chip market, Japan held a 95 percent share of the market in 1988. With the advent of memory-hungry VTRs, televisions, and computers, the Japanese have carved out a formidable position in the marketplace.

In retrospect, it can be argued that the VLSI Project would have succeeded without Nebashi's taking his team out after hours, but breaking down the barriers did help unleash an extra surge of creative thinking among the member companies. MITI has now adopted many management ideas from the VLSI Project for its next-generation programs. Getting rival companies to share generic technologies and creative ideas over a few drinks is clearly one of them.

Designing Creativity-Inducing Environments

Creating a sense of camaraderie outside the office or research laboratory is only a partial solution to the stifling working environment that most Japanese face every day. In recent years, creating an environment conducive to original thinking—or *kankyo-zukuri*—has become an industry in its own right because of the trend toward brain-intensive work.

For years, Japanese companies have boasted efficient, spotless factories, but their offices resemble open bull pens overflowing with papers and files. For anyone trying to do any creative work, most Japanese offices are too noisy and chaotic, making it virtually impossible to concentrate or think very deeply. Telephone calls, visitors, and impromptu visits from customers are constant interruptions, so most Japanese wait until evening to get any serious work done. The obvious drawback to working late nights, however, is that productivity and creativity drop off sharply because many Japanese office workers suffer from extreme stress and exhaustion. Thus, improved office environments are a way to reduce the need for perpetual after-hours socializing and working.

Japan's new breed of young people—better known as *shinjinrui*—want better office layouts, carpeting, and wall partitions to reduce interruptions and allow for more individuality. Companies such as Ricoh and Sanyo are building Silicon Valley–style offices and cubicles for their researchers. Office interior design company Okamura sells new petal-shaped desks and open partitions, while Kokuyo markets "Creation Live" office furniture designed to promote creativity. Meditation rooms—quiet study rooms for escaping to from the hustle and bustle of the office and, for some, a place to take a quick nap—are gradually being introduced. Japanese construction companies are capitalizing on the demand for "intelligent" buildings and office complexes by offering ergonomically designed offices, landscaped views, contemplative art, and muted colors to put employees at ease. The goal is to create a mood for more creative research and more productive human interaction.

The Ministry of Construction estimates that the "intelligent building" market will generate $300 billion in business for construction, electronics, and service companies over the next ten to fifteen years. With these huge sums of money being spent on improved facilities, Japanese architects and landscapers are rapidly becoming world leaders in the design of next-generation office environments. In the near future, Japanese office design could well become a key export industry. Japanese investors already play a major role in U.S., European, and Asian real estate markets.

A pacesetter in creative office interiors is stationery goods maker Plus and Company, whose president Yoshihisa Imaizumi has created a relaxed, spontaneous office environment. In the *Journal of Japanese Trade and Industry,* Masahi Kurata observes:

> The atmosphere of the [Plus] office is strikingly different from that of most Japanese company offices. Employees are free to dress as they like. There is none of the compulsory dark blue or gray suits for men and uniforms for women at Plus. Nor is there a single one of the drab, gray steel desks or chairs found at other companies. Instead, the office as a whole is laid out to leave ample space for individual staff members, without obstructing interpersonal communication. . . . Desks for work requiring mental concentration . . . office automation equipment operation and drawing, for instance, are lined with partitions about 1.5 meters high. . . . But lower partitions are installed in product planning and other offices where freewheeling internal communication is essential. . . . Unique for Japan, there are tea corners with little round tables and coffee makers scattered around here and there in the office.

Although commonplace in the West, Plus's informal approach is unique in Japan and has helped the company develop original products. In 1986 Plus developed miniature writing kits and the "Team" mini-stationery set, two instant hits in Japan. Its Copy-Jack photocopier, developed with Kyushu Matsushita Electric, created an entirely new market for electronic blackboards.

In Japan's competitive marketplace, designing office environments conducive to creative thinking will spell the difference between success and failure for many companies.

Venture Business Incubators

Entrepreneurialism and venture capital are two economic phenomena that have filtered into Japan over the last ten years. In the early 1980s start-up companies—known as "venture businesses" in Japan—blossomed as a result of Japan's emerging venture capital industry and booming export business. Hundreds of small computer and software houses appeared in Tokyo, Nagoya, and Osaka in the shadow of large conglomerates. They were joined by over forty new venture capital firms.

After several years of heady growth, Japan's venture businesses hit a serious downturn when the yen shock sent exports plummeting in 1985. According to the *Japan Economic Journal,* these venture capital firms invested about $1.6 billion into 3,425 venture businesses in 1987, a substantial increase, but despite an infusion of venture capital from Japanese investors, it is still difficult for new venture businesses with creative ideas to secure financing, talented people, and customers. Japanese industry is still dominated by vertically integrated corporations that belong to large financial groups (*keiretsu*) with extensive ties to bankers, suppliers, distribution channels, service outlets, and government ministries. Moreover, Japan's conglomerates can outspend venture businesses in research and marketing.

As venture capital analyst Yoriko Kishimoto notes:

> In the venture capital boom of 1982–1985, a number of quite young and worthwhile technology companies were flooded with venture capital, but some of them only managed to speed ahead to spectacular bankruptcies. Although they finally had access to capital, the Japanese system still denied them their own lifelines: a fluid and supportive labor market, a customer base receptive to new companies' products, and a qualified management base. The employment system is still rigid and slanted towards the large, public companies.

In Japan's risk-averse environment, after-hours socializing

and improving the work environment may be useful for stimulating the creativity of "intrapreneurs" in existing organizations. But these methods are inadequate for venture business entrepreneurs, especially those operating outside of the corporate network of services, talented labor, financing, and distribution.

To provide support for budding entrepreneurs, venture business associations and business incubators are being established throughout Japan. In the Tokyo region, Kanagawa Prefecture has taken the lead by creating the Kanagawa Science Park, the Kawasaki Microcomputer City, the Yokosuka Research Park, and the Yokohama Business Park, where numerous computer, software, and telecommunications venture businesses are receiving seed financing, management consulting, and marketing advice. The city of Kyoto, a traditional seedbed for start-up companies, has actively supported venture businesses. In 1985 the Kyoto Venture Business Club was formed. In April 1989 Kyoto City opened a venture business incubator to coordinate research in mechatronics, computer-aided design, and information services.

At the national level, MITI has developed a variety of incubator programs for high-risk high-technology ventures. These stimulative programs include the Venture Enterprise Center, the technopolis program, and the regional research core program and are aimed at "jump-starting" entrepreneurial companies into operation.

Venture Enterprise Center

In 1975 MITI established the Venture Enterprise Center (VEC) to guarantee low-cost loans to small businesses for R&D activities. During the venture boom of 1982 to 1985, MITI downplayed its role and emphasized the role of venture capital companies, since most Japanese entrepreneurs shied away from the paperwork and MITI interference. By March 1988, VEC had guaranteed over 500 loans. As supplementary programs, MITI initiated four venture business loan programs in 1984: the Small Business Finance Bank program, an R&D subsidy program, subsidies to new ceramic research projects, and subsidies to local industrial testing laboratories for purchasing computers. These programs, designed to help venture businesses underwrite research costs, are part of MITI's policies to promote infant industry.

The venture businesses funded through VEC and MITI's other programs have had mixed success. Most Japanese entrepreneurs still prefer self-financing and loans from family and friends to maintain confidentiality and control over their operations.

Technopolis Program

Realizing that Tokyo's hyper-competitiveness has contributed to Japan's imitative technology and that, ultimately, a specially designed environment will be required to support extensive creative research, MITI has begun constructing twenty-six research cities—or technopolises—throughout the country. Based on Silicon Valley, the Tsukuba Science City, and the Sophia Antipolis in southern France, the technopolises are envisioned as incubators for scientific and technological research in the twenty-first century. They will feature research universities, techno-centers, research parks, joint R&D consortia, venture business incubators, intelligent office complexes, international convention centers, and residential new towns.

Under the Technopolis Law, passed by the Japanese Diet in April 1983, twenty-six technopolis areas (see Figure 7–1) were chosen from thirty-eight candidate sites because they met five criteria: (1) they integrated the development of industry, research universities, and housing; (2) close ties existed between the technopolis and its "mother city" of 200,000 or more people; (3) development was balanced between high-tech industries and local industries; (4) both transfer of high technology from Tokyo and Osaka to the regions (transfer R&D) and creative research in frontier fields (frontier R&D) would be possible; and (5) regional high-tech research was unique. Although the technopolises qualify for national tax incentives, special depreciation allowances, low-interest loans, and plant siting assist-

FIGURE 7–1
Japanese Technopolis Areas

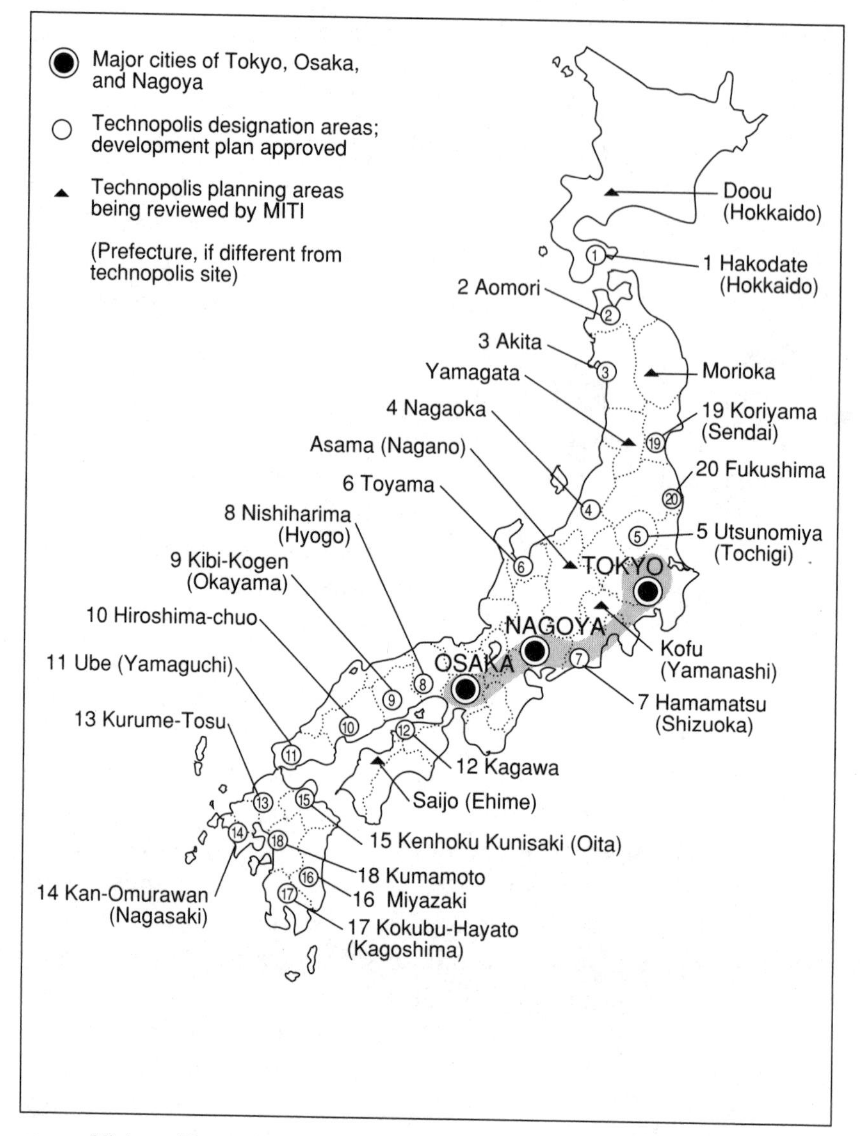

SOURCE: Ministry of International Trade and Industry.

ance, MITI has delegated the actual responsibility of planning and financing the technopolises to local governments. MITI acts primarily as a catalyst and program coordinator.

Since 1984 the technopolises have prepared twenty-year

development plans, organized over 270 R&D consortia, and begun constructing highways, airports, industrial parks, and new towns according to plan. Twenty of the technopolises have begun "Technomart" programs—information distribution centers—to provide on-line data bases to users in the shipping, distribution, marketing, tourist, and retail industries.

The infrastructure to support a future generation of researchers is gradually taking shape in the technopolises. But global economic shifts have undermined the assumptions behind the technopolis program. Since the yen shock of 1985, Japanese manufacturing plants have moved offshore to less expensive sites in Asia, Europe, and the United States, reducing the flow of technology from Tokyo and Osaka. In 1986 plant siting in the technopolises took a sharp nosedive, though it recovered somewhat in 1988. Another setback was that regional governments initially focused their efforts on "hard" infrastructure projects, such as roads, airports, and highways, and underestimated the difficulty of developing the "soft" infrastructure of R&D consortia, venture capital funds, and university research needed to drive the technopolises. Most technopolises are still empty shells, and top-flight Japanese engineers still prefer to live and work in the Tokyo area, whose wealth of educational and cultural resources attracts 80 percent of the nation's researchers. Unlike Tokyo or Silicon Valley, the technopolises are not beneficiaries of a natural flow of people and jobs.

Professor Kenichi Imai of Hitotsubashi University, a member of MITI's Technopolis '90 Steering Committee, believes the technopolis program might achieve moderate success but it may not be the right solution to Japan's quest for creative research.

> The goals of the technopolis policy are to promote manufacturing-type R&D through technology transfer from Tokyo and Osaka, and to establish locations where knowledge-oriented R&D will be undertaken. . . . Although it is too early to evaluate the policy, implementation of [manufacturing-type] R&D appears to be progressing at a steady pace . . . [but] it would

be overoptimistic to expect new technological breakthroughs from these projects.

Regional Research Core Program

To overcome the many obstacles facing the technopolises, MITI developed the "regional research core" concept in 1986 to provide funding for specific research activities. The program promotes four types of research incubator facilities: joint industry-university-government research institutes, new research training centers, venture business incubators, and international conference and exhibition halls. MITI is also exploring intelligent office complexes that will house venture capital firms, legal and accounting professionals, think tanks, and data services. The private sector is being encouraged to develop research core projects according to MITI guidelines and will be eligible for tax benefits, insurance guarantees, and loans from the Japan Development Bank (JDB) and Hokkaido-Tohoku Development Corporation (see Figure 7–2).

As of early 1989 MITI had designated twenty-eight cities as research core cities (see Table 7–1) to conduct research on biotechnology, electronics, communications, new materials, and energy. Many of these cities are technopolis sites. The major difference between this program and the technopolis program is that technopolis is more manufacturing-oriented, while the regional research core program focuses more on research and services.

The 21st Century Plaza in Miyagi Prefecture north of Tokyo, for instance, is a $213 million complex featuring two zones: a "techo-culture zone" of research institutes, information centers, venture business incubators, computer shops, and an international convention center zone. The Technovalley Intelligent Core in Niigata Prefecture on the Japan Sea coast is in the Nagaoka Techno-Valley Technopolis, an advantageous locale. This $100 million project will feature the Nagaoka Regional Technology Development Promotion Center, the International Communication Plaza, an education plaza, and venture business incubators. Because Nagaoka is located only

FIGURE 7–2
The Regional Research Core Program

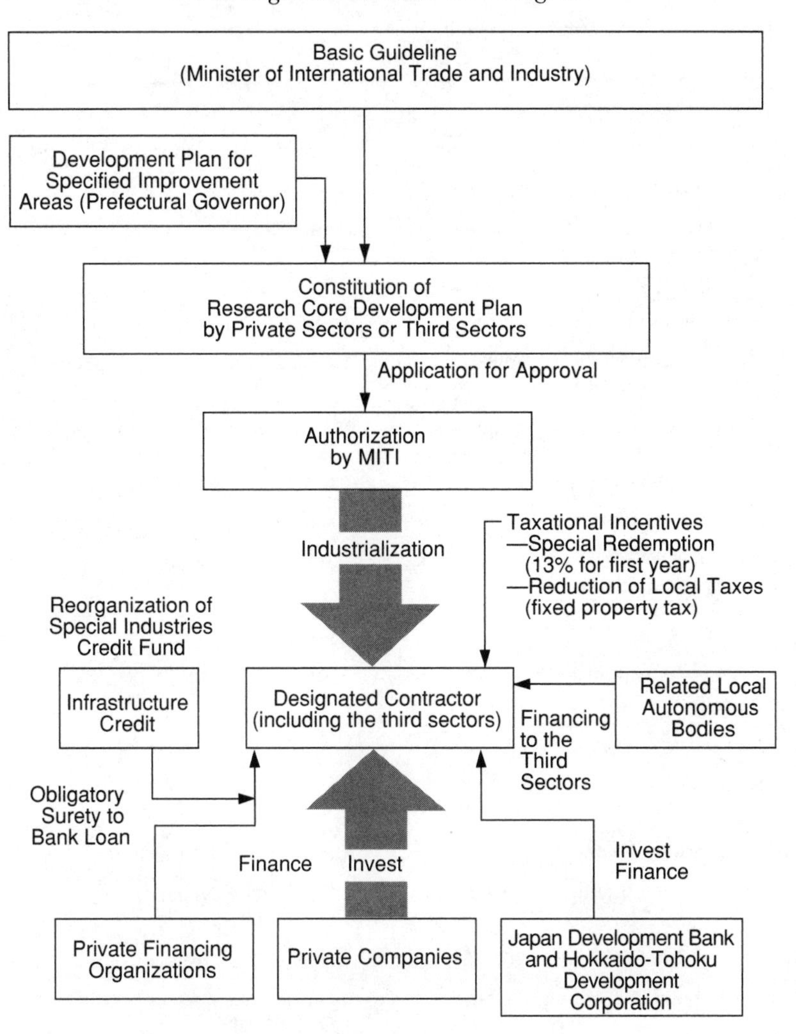

SOURCE: Ministry of International Trade and Industry.

two hours from Tokyo on the Joetsu bullet train, some Tokyo businesses are already moving there to escape escalating land costs.

MITI is working with private industry to expand the technopolis and research core programs into a nationwide system of twenty-first–century research cities linked by computer networks. This program is MITI's vision for cultivating researchers

TABLE 7-1

Regional Research Core Projects

Name of Project	Location
1. Eniwa High Complex City	Eniwa City, Hokkaido
2. Aomori Future Park	Aomori City, Aomori Pref.
3. Technopolis Support Core	Morioka City, Iwate Pref.
4. Akita Core City Key Area Industrial Support Functions Development Project	Akita City, Akita Pref.
5. 21st Century Plaza	Izumi City, Miyagi Pref.
6. Yamagata International Industrial Communication Plaza	Yamagata City, Yamagata Pref.
7. Koriyama Regional Technopolis Industrial Support Functions Development Project	Koriyama City, Fukushima Pref.
8. Technopolis Center	Utsunomiya City, Tochigi Pref.
9. Hitachi-Naka International Bay Park City "Business Pleasure Hitachi-Naka"	Katsuta City, Ibaragi Pref.
10. Kazusa New Development and Research City Plan "Academia Park"	Kisarazu City, Chiba Pref.
11. Kanagawa Science Park	Kawasaki City, Kanagawa Pref.
12. Science Park (21st Century Industrial Park)	Kofu City, Yamanashi Pref.
13. Technoculture Zone	Nagano City, Nagano Pref.
14. Technovalley Intelligent Core Project Plan	Nagaoka City, Niigata Pref.
15. Toyama Intelligence Corridor	Toyama City, Toyama Pref.
16. Cosmopolis Plan (Research Park)	Kishiwada, Izumi-Sano, Izumi cities, Osaka Pref.
17. Okayama Triangle R&D Project Plan	Okayama City, Okayama Pref.
18. Hiroshima Central Technopolis Innovation Park	Higashi-Hiroshima City, Hiroshima Pref.
19. Ube New City Techno Center	Ube City, Yamaguchi Pref.
20. Tokushima Prefectural Industrial Research Core	Tokushima City, Tokushima Pref.

TABLE 7-1 (*continued*)

Name of Project	Location
21. Techno-Plaza Ehime	Matsuyama City, Zhime Pref.
22. Kurume-Tosu Research Park	Kurume City, Fukuoka Pref.
23. Saga Research Core	Saga City, Saga Pref.
24. Creative Area Project	Nagasaki City, Nagasaki Pref.
25. Kumamoto Techno-Creative Area	Eijo District, Kumamoto Pref.
26. Oita Intelligent Zone	Oita City, Oita Pref.
27. Miyazaki Sun-Tech Park	Miyazaki City, Miyazaki Pref.
28. New City Center	Hayato Town, Kagoshima Pref.

SOURCE: Ministry of International Trade and Industry.

for Japan's next-generation industries, such as biotechnology, new materials, aerospace, electronics, and superconductors. As the saying goes, however, "Rome was not built in a day." MITI is patiently waiting for its technopolises and research core programs to take root.

Thus, the Japanese are pursuing new approaches to cultivate creative ideas. Next, let's look at methods being developed to generate new breakthrough ideas and technologies.

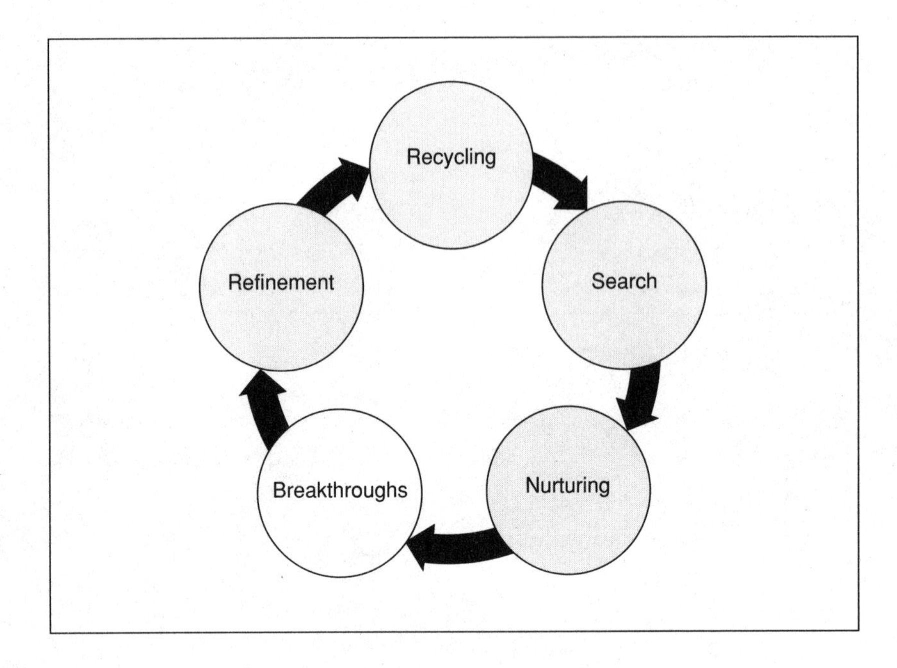
Recycling
Search
Nurturing
Breakthroughs
Refinement

発想
HASSOO

8

Hassoo: Generating Breakthroughs

> The next revolutionary advance in technology should come by the beginning of the next century, most probably from Japan.
>
> —Dr. Michiyuki Uenohara
> Executive Vice President, NEC Corporation

THE Japanese are expert at improving, refining, and recycling ideas, but rarely have they made the spectacular discovery or product breakthrough that has led to an entirely new industry. Most innovations coming out of Japan today, such as videophones, fiber-optic networks, and high-temperature superconductivity, are rooted in scientific theories and ideas developed in the West. Although technology fusion—the synergy of brilliant minds—is now being used by the Japanese to tap their inner resources, this creativity-inducing method actually stifles Japan's most creative minds, a situation that is seriously handicapping Japanese industry in global competition. Clearly, other approaches need to be found.

The mood in Tokyo is changing. Once viewed as a threat to the status quo, breakthrough ideas are no longer taboo. Companies are openly seeking ways to foster breakthrough thinking among their employees. They are sending people around the world to find new breakthrough techniques and looking inward for new methods of their own. This search is leading them through a major industrial reeducation process not unlike their search for quality decades ago.

Hiroshi Takahashi, for example, has run creativity training courses for production managers and scriptwriters at the state-run Japan Broadcasting Corporation (NHK) since 1965. Recently, he published a handbook on creativity training that codifies the lessons he has taught over the last twenty years. He observes that Japanese companies are now experimenting with over 100 creativity techniques. Half of these methods were introduced from the United States during the 1950s and 1960s, while the other half are new methods being developed in Japan. A. F. Osborne's "brainstorming" (1953) and W. J. J. Gordon's "synectics" (1961), for example, are two ideas that have taken root in Japan. By contrast, the Japanese have developed a variety of brainstorming techniques that take advantage of Japanese group dynamics. Although these creativity tools are still being tested, they give one a fascinating glimpse into Japan's next move in science and technology—the leap to creative research.

The KJ Method

One of the early postwar pioneers in Japanese creativity research is Jiro Kawakita, a cultural anthropology professor formerly at the Tokyo Institute of Technology and now at Chubu University. In 1964 Professor Kawakita developed the "scrap paper method" (*kami-kire ho*), a method for generating new conceptual "images" from raw data. Professor Kawakita uses the KJ (Kawakita Jiro) method in exploratory research that has no preexisting theories or conceptual "handlebars" to hang onto.

The KJ method is conducted in four steps. In the first step, participants are given a theme and are then asked to write down as many ideas as possible onto small cards (which have replaced the original scrap paper), limiting their ideas to twenty to thirty

Japanese characters (*kanji*) per card.* The goal is to generate 100 cards or more. (When Professor Kawakita conducted field research in Nepal and the Himalayas, he generated over 800 note cards.)

The second step is to organize the cards into categories of 50–100 cards. The cards are sorted again into groups of 20–30 cards according to common themes, then sorted into smaller groups of ten cards or less. Professor Kawakita encourages a playful spirit when sorting the cards into new conceptual frameworks. By breaking large groups of ideas up into smaller groups, the KJ method attempts to break down rigid thinking and open the door to new ideas.

The third step is to sit back with a large sheet of paper, relax, and write down ideas that come to mind. The ideas may be related or unrelated; they may be different perspectives on the problem at hand. They are graphically depicted to make it easy for the participants to quickly understand their meaning and relationship to one another. By the end of this exercise a conceptual picture of the new ideas emerges for all to see.

The final step is to read aloud groups of ideas on the conceptual picture and write down new ideas that are triggered by the picture or the discussion.

The KJ method, like other Japanese creativity techniques, uses complex associations among ideas as a way of triggering new ideas; it is widely used among Japanese researchers and managers. In 1965 the Japanese Creativity Association published a book describing Professor Kawakita's method, which has become popular among members of the Japan Management Association. Today, it is common to see conceptual road maps, or images, in many government and corporate reports. For example, in his ground-breaking book on Japan's future telecommunications network, Dr. Yasusada Kitahara of Nippon

*Many Japanese creativity techniques employ cards and the skills developed from playing a traditional Japanese card game—*karuta*—which is based on *carta*, a Portuguese game introduced to Japan in the sixteenth century. *Karuta* involves matching the opening lines of 100 famous thirteen-line poems (*tanka*) from the thirteenth century (*Hyakunin Isshu*) with their correct endings. The game teaches Japanese the ability to memorize and draw associations among the pictures of princes and ladies on the playing cards.

Telegraph and Telephone (NTT) developed a twenty-year technology road map for NTT's future digital information network system (INS) to show the conceptual merging of telephone, facsimile, video, and data communications. Indeed, compared to Westerners, Japanese tend to think more visually, one of the reasons the KJ method has been so influential.

Creativity Circles

Harnessing the creativity of Japanese brainstorming has become a new challenge for many leading Japanese companies in recent years. And not surprisingly, they are looking to the quality circle (QC)—one of the secrets of Japan's phenomenal economic success—as a vehicle for promoting creativity. Unbeknownst to many Westerners, Japanese companies are using what have now evolved into "creativity circles" to experiment with a variety of brainstorming techniques.

How do these creativity circles work? In Japanese companies groups of engineering, marketing, sales, and manufacturing employees are gathered to discuss not only problems with existing products but also ways to develop new "hit" products. These product development teams often consist of five people (*gonin-gumi*)—the number of people on a team considered ideal by the Japanese because it allows for both flexibility and a variety of opinions. The brainstorming is incremental in nature and focused on the immediate task at hand. Initially, the team members are asked to suggest refinements to existing products, such as using fewer manufacturing steps or components. Over time, the teams turn to creative solutions by implementing some of the brainstorming techniques described earlier.

The team inevitably encounters obstacles along the way that it is unable to solve using conventional QC techniques. At this point each member is encouraged to personally think up totally new approaches—after hours over drinks, on the train home, or on weekends. Members present their ideas to the creativity circle, where they are analyzed for their potential use in solving the problem under discussion.

Yamaha's digital sound-field processor, one of the 1986 winners of the Nikkei Annual Award for Creative Excellence, is an example of Japanese creativity circles in action. In early 1984 Yamaha pulled together a team of researchers from its technical laboratory, electronic instruments division, and architectural acoustic research section to develop a new type of digital sound recording. Its goal was to replicate the rich sonic "afterglow" that results from the early reflections and later echoes off chamber walls. The development team recorded acoustical data from concert halls, jazz clubs, and cathedrals onto a ROM chip. Parlaying the varied technical background of its members, the team simulated ideal sound fields using CAD techniques, conducted tests on a variety of music performances, and then designed the stored data into a VLSI chip. The resulting product—the Yamaha DSP-1—was introduced in Yamaha stereo equipment in June 1986 and astonished music listeners with its ability to create spatially realistic sounds for sixteen different performance spaces.

Multiple-Track Development

In any industrial or retail sector in Japan, there are likely to be a dozen competitors battling it out for market share. Akihabara—a flourishing electronics bazaar in Tokyo that began as a postwar black market—encapsulates the fierce competition among companies. During an average day, thousands of hobbyists and businessmen pour through the twelve-block area searching for electronic parts and systems. Akihabara exhibits the latest in Japanese electronics wizardry, most of which will neither survive domestic competition nor be exported abroad. Displayed in rows are dozens of color TVs, calculators, video equipment, computers, and satellite antennas priced within several dollars of each other. Each product offers a slightly different feature that vendors hope will attract customers. Each product also represents years of hard work by engineering and marketing talent in the competing companies.

Because of the fierce competition, both domestic and foreign, Japanese companies are reluctant to pin their hopes on one technology or product. Thus, they will fund several groups, both to stimulate internal competition and to devise fallback positions in case one group runs into technical dead ends. In MITI's Supercomputer Project, for example, Fujitsu, Hitachi, and NEC are simultaneously pursuing a variety of ultrafast chip technologies—gallium arsenide, emitter-coupled logic (ECL), Josephson junctions, and superconductors—to build their next-generation supercomputers. These companies are seeking advances in multiple areas and are wary of "putting their eggs in one basket" because there may be unforeseen breakthroughs, such as IBM Zurich's high-temperature superconductors.

Indeed, multiple-track development is essentially a conservative research strategy for cash-rich, risk-adverse companies. Even parallel (two-track) development is vulnerable to technical, legal, and political risks. During the early 1980s, for example, NEC pursued a parallel-track approach in developing its microprocessors. It licensed Intel's microprocessor technology, while developing its own proprietary V series microprocessors to establish itself in the market. The underlying rationale for its parallel-track development was to ensure continuity in the event that Intel did not renew its license.

In 1984 Intel terminated its second-sourcing licensing and took NEC to court for reverse engineering its microprocessors. While the legal suit dragged on in court, NEC was forced to develop a new microprocessor, the V33, which was significantly different from Intel's 8086 and its own proprietary V30 series. The Intel-NEC suit illustrates the dangers of relying on parallel-track development and the advantages of multiple-track (three or more) development. Two development tracks can be delayed for a variety of reasons. Without multiple-track R&D, Japanese companies dependent on Western technology can easily be stopped by lawsuits by American companies seeking to protect their copyrights and patents.

Brainstorming

Most Japanese creativity methods are rooted in the Western concept of brainstorming because of the primacy of the group in Japanese society. In traditional brainstorming sessions, the ground rules are simple enough: a topic is chosen and everyone is encouraged to think of as many ideas as they can, no matter how ridiculous or outlandish, for a half hour or longer. The quantity of ideas, not their quality or usefulness, is the main goal of the exercise. To avoid "group-think," criticism and value judgments are discouraged until the session ends or the group runs out of ideas.

Although brainstorming is useful for generating new ideas, it often leads to highly impractical ones because the participants feel no responsibility for the ideas they express, nor do the sessions encourage participants to develop and refine their ideas. Although brainstorming works well when the participants are vocal and individualistic, to the Japanese, Western-style brainstorming is like throwing clay haphazardly onto a wheel without molding it into fine pottery: it is unfinished and incomplete. In Japan, unstructured brainstorming sessions often draw blank stares because the Japanese tend to defer to the opinions of others. They realize that ideas carelessly tossed out have the power to embarrass and ridicule. Thus, Western-style brainstorming often does not work in Japan.

The Mitsubishi Brainstorming Method

At Mitsubishi Resin, Sadami Aoki has developed an alternative, the Mitsubishi brainstorming (MBS) method, to take advantage of the Japanese preference for structure and order. In this approach, the session lasts longer—up to two or three hours—to allow the participants to "break the ice" and develop ideas more fully. Rather than openly voice their ideas at random, the participants are given a chance to warm up by writing their ideas down for fifteen minutes. Then, to prevent more aggressive or vocal people from dominating the group, each person is asked to read his or her ideas aloud. Everyone is encouraged to write down new ideas as they listen to those of other participants, thus enabling those initially without ideas to "save face." For the next hour or so, the participants are asked to explain in greater detail the background and content of their ideas, which are written onto "idea maps" by the group leader. In this more structured, organized way, the Mitsubishi brainstorming method attempts to elicit ideas from all participants.

The NHK Brainstorming Method

Hiroshi Takahashi, the author of *The Creativity Handbook,* has developed another brainstorming method from his years of training television production managers at the Japan Broadcasting Corporation (NHK). In the NHK brainstorming method, the participants are asked to write down five ideas (one per card), then gather into groups of five people. While each person explains his or her idea, the others write down new ideas that come to mind. Then the cards are collected, sorted, lined up, and grouped into related themes. New groups of two and three people are formed, and the themes are brainstormed for half an hour while the participants write down new ideas on cards. At the end of the session, each group organizes the new ideas by theme and announces them to the larger group. All of the ideas are written on a blackboard. The participants are then reorganized into groups of ten people, and the ideas on the blackboard are brainstormed again, this time one idea at a time. Although rather long and complicated, the NHK brainstorming method works like an egg beater, churning ideas again and again until a new mixture is generated.

The Lotus Blossom Technique

Recently, the Japanese have returned to their cultural and religious origins to develop new creativity techniques. One of the more original is the MY (Matsumura Yasuo) Method, or lotus blossom technique, which was developed and patented by Yasuo Matsumura, president of Clover Management Research in Chiba City. The MY method is used by the Japan Management Consultants Association (Nihon Keiei Gorika Kyokai) to train managers in creative brainstorming and management planning design. This approach is not completely foreign to Westerners because it also forms the key concept behind Lotus 1-2-3, the famous spreadsheet software program developed by Mitch Kapor, founder of Lotus Development Corporation and long-time student of Eastern religion and philosophy. In the MY Method, a core theme is pursued into ever-widening circles of petals, or "windows." The only difference is that in Lotus 1-2-3 the petals are lined up as windows along the top of the computer screen.

How is the lotus blossom method used? As shown in Figure 8–1, a central theme is written in the center of the lotus blossom diagram and participants are asked to think of related ideas or applications of the idea. For example, if the theme is superconductivity, participants might think about commercial applications, such as magnetic levitation trains, energy storage, electrical transmission, computer board wiring, and so forth. These applications are written in the surrounding circles labeled A to H in the center box.

In the next step, these ideas become the core themes for the surrounding lotus blossom petals or boxes. For example, if "electrical transmission" is written in circle A, then it is the core theme for the lower middle box. Participants are then asked to think of eight new applications for superconductors in electrical transmission, such as high-voltage lines and computer wiring. These ideas are written in the surrounding circles marked 1 to 8. When the themes become too specialized or technical, they can be brainstormed by groups of experts. Product planning groups can take the lotus blossom diagrams to divisions specializing in different aspects of the new idea for further elaboration. This process can be repeated again and

FIGURE 8–1
The Lotus Blossom Method Diagram

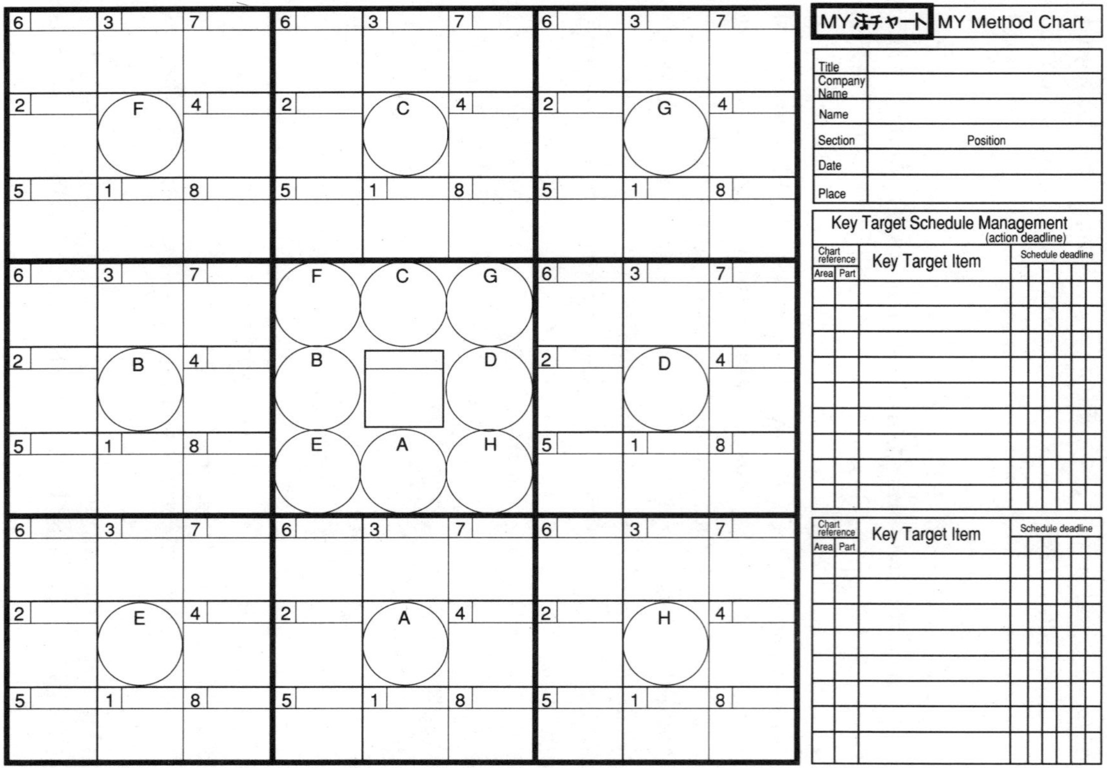

6 3 7
2 F 4
5 1 8

6 3 7
2 C 4
5 1 8

6 3 7
2 G 4
5 1 8

6 3 7
2 B 4
5 1 8

F C G
B D
E A H

6 3 7
2 D 4
5 1 8

6 3 7
2 E 4
5 1 8

6 3 7
2 A 4
5 1 8

6 3 7
2 H 4
5 1 8

MY法チャート MY Method Chart

Title
Company Name
Name
Section
Position
Date
Place

Key Target Schedule Management
(action deadline)

Chart reference		Key Target Item	Schedule deadline
Area	Part		

Chart reference		Key Target Item	Schedule deadline
Area	Part		

SOURCE: Clover Management Research, Tokyo, Japan. Reprinted by permission.

again until hundreds or thousands of new ideas are generated from the original idea.

The Japanese are well known for their superb ability to generate commercial applications through their use of technology trees and their own forms of brainstorming. The lotus blossom (MY) method is probably the most effective because of its simple, open-ended structure that allows for systematically linking new ideas.

According to Shigeo Ihori, president of the Tokyo software house Taiyo Kikaku:

> When I was at the Japan Productivity Center, we used the MY method to help companies diversify into new products and technologies. It was useful for helping businessmen structure their thinking into creative new avenues. You might even say that the lotus blossom method is a Japanese form of brainstorming.

The Feed-Forward Method

A familiar name to many Japanese is Kenichi Takemura, a social critic and former Fulbright scholar at Yale and Syracuse universities; his book *The Era of Genius* (*Isai Jidai*) argues that the changing global economy will require greater creativity and genius in Japan. Based on his experience, Takemura offers a variety of creative approaches, including "feed-forward" thinking, multiple-activity thinking, reverse (contrarian) thinking, "amoeba" (multi-minded, as opposed to single-minded) thinking, and "45-degree" (peripheral vision) thinking. Perhaps the most interesting is his feed-forward method.

Takemura observes that most systems are based on feed*back* methods, not feed-forward methods. For example, room air conditioners are designed to pump cool air only after the temperature in the room has risen above a certain level. But responding to changes is inefficient and often happens too late;

even if the air conditioner works properly, it must recool heated air. In a feed-forward approach, the system would respond before undesirable external conditions trigger the system into action. Air conditioners could be equipped with atmospheric pressure sensors or sunlight sensors that would automatically adjust the air conditioner in anticipation of warming temperatures. By anticipating changes in the environment, the system would operate in a more efficient and timely manner.

In electronics, many Japanese companies have adopted a forward-looking or anticipatory attitude (*maemuki shisei*), which is one kind of feed-forward practice. Rather than waiting for competitors to take the lead, they are anticipating change and adjusting their corporate strategies accordingly. Japanese television makers such as Sony, NEC, and Toshiba have invested heavily in HDTV and VCR technologies because they expect large markets in the future. Rather than invest in the future of these technologies, many U.S. companies, on the other hand, are waiting for the market to develop because of their short-term concern over negative feedback from Wall Street. Unfortunately, the long development time required for HDTV and other emerging products reduces the chances of success for late entrants. In business, the element of surprise is the best strategy. By responding to events, U.S. companies drastically reduce their strategic options over time. Indeed, the paucity of investments by large U.S. companies in HDTV bespeaks our reactive feedback mentality and has already put our companies on the defensive. In the future, victory will go to feed-forward companies.

In this chapter, we have reviewed some of the breakthrough techniques being experimented with in Japan. Creative research is still in the early stages for many companies, so there are few conclusive results. Indeed, it can be argued that Japanese companies have made few breakthroughs because they have largely responded to Western breakthroughs in the past. Now that they are being prodded by the West to pursue their own proprietary technologies, we may see Japanese breakthroughs in the 1990s. Until Japan strengthens its basic research facilities, however, major breakthroughs are likely to come from "hit" products, not scientific discoveries, for some time to come.

In Chapter 13, we will look at how Japanese universities, government ministries, and corporations are trying to promote creativity in basic scientific research.

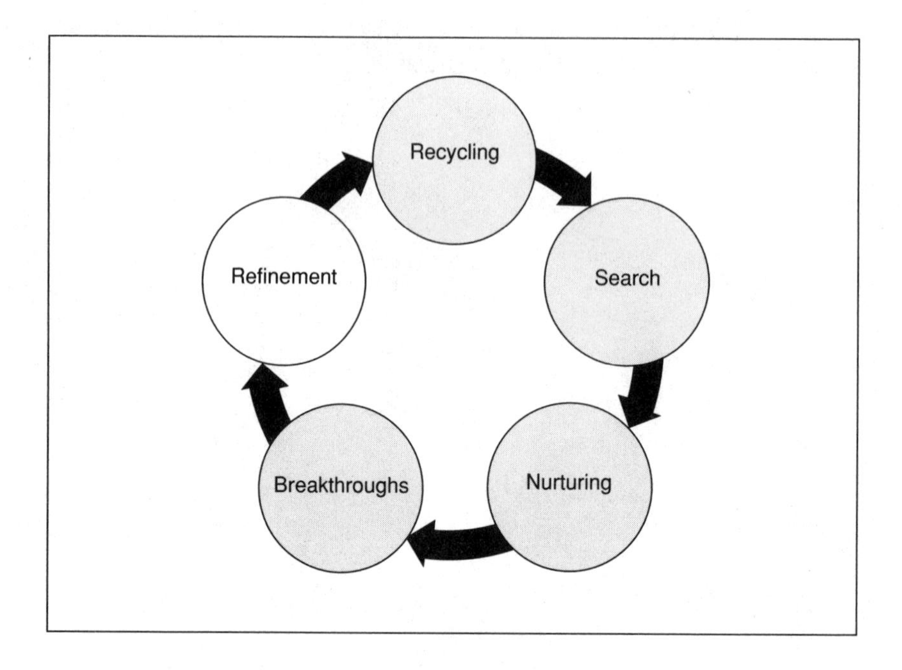

KAIZEN

9

Kaizen: Refining Ideas

> When it comes to opening completely new fields, our scientists are still catching up. . . . However, when it comes to turning ideas into reality, we have been very successful and very creative.
>
> —Akio Morita
> *Made in Japan*

OF all the creative methods of the mandala, idea refinement is Japan's greatest strength. In consumer electronics and computing, Japanese companies demonstrate a tremendous knack for refining new ideas and technologies. They often take seminal ideas overlooked or dismissed by skeptical Westerners and gradually transform them into something entirely different. The videotape recorder (VTR) is a classic example of Japan's genius for creative refinement. Although Ampex pioneered VTR technology for professional systems, it was Sony and Matsushita that turned the VTR into a mass consumer product by selling a new marketing concept, time-shifting—the recording of television programs for delayed viewing. What was originally a professional tool suddenly became a means for freeing television viewers from the tyranny of broadcasting schedules.

Contrary to popular belief, the creative refinement of the Japanese is not merely product imitation or copying. It is a disciplined method for transforming an idea into something new and valuable, just as cutting and polishing diamonds is an industrial process for creating jewelry. While innovation involves making small improvements in existing products, such as better VTRs and faster computers, creative refinement introduces totally new marketing concepts and applications. It is conceptual leapfrog. Sony, for example, redefined the concept of portable sound by using stereo headphones to develop the Walkman, creating a whole new way of listening to music. No longer do listeners have to sit in an auditorium or living room to enjoy stereo sound. They can bring it with them shopping, jogging, or lounging at the beach. Similarly, Japanese video makers have transformed home movies into an interactive cinematographic experience. By offering superior audiovisual quality and instant feedback, Japanese 8mm video cameras have made Shakespeare's famous line—"All the world's a stage"—come alive at many family get-togethers and special events.

The concept of *kaizen*—or constant improvement—is central to understanding Japan's approach to creative refinement. Unlike Western companies, which often seek quantum leaps in technology and productivity, Japanese companies prefer a series of small, incremental improvements in their daily operations. They hit for singles and doubles, not the home run or grand slam. Japanese workers are constantly encouraged through the use of quality circles, suggestion boxes, meetings, and consensus decisionmaking (*ringi*) to offer new ideas for reducing costs, increasing quality, simplifying procedures, and developing new products. They are taught that small improvements, no matter how minor or insignificant, are crucial for the ultimate success of their company. If there is any secret to Japan's economic successes, *kaizen* surely ranks high on the list.

Today, the concept of *kaizen* is being adapted to basic research and new technologies. Unlike in the past, it is now being used to creatively refine ideas, not just products. In the field of personal computing, for example, Toshiba and other computer makers are rapidly introducing generation after generation of new ideas into their laptop computers. Initially, they introduced liquid crystal display (LCD) screens and built-in printers, then memory storage cards, optical scanners, and ergonomic keyboards, and now voice recognition. In the field of high-temperature superconductivity, Japanese researchers are refining new theories and experimenting with various materials in hopes of uncovering new breakthroughs.

What are some methods commonly used by Japanese companies to refine new ideas and technologies? How are they used to transform ideas into innovative products and services? This chapter reviews five powerful Japanese techniques: miniaturization, simplicity, incompleteness, visualization, and transformation.

Miniaturization

The Japanese are well known for their proclivity for miniaturization, which is reflected in their compact, lightweight radios, cameras, televisions, and automobiles. This expertise in miniaturization is highly developed, even when compared with other Asian nations. The Chinese, for example, are master carvers of miniature jade and ivory figurines, but unlike the Japanese, they did not incorporate the idea of smallness into their very culture.

Korean writer O-Young Lee, author of *Smaller Is Better: Japan's Mastery of the Miniature,* observes that the Japanese predilection for smallness is not found even in nearby Korea, which was the source of many Japanese traditions.

> In Japanese . . . it is not the prefixes for "huge" but those for "tiny" that are commonly used. . . . The round bean (*mame*) is like a condensed representation of the world. . . . Such words as *mame-bon* (mini-book), *mame-jidosha* (mini-car), *mame-ningyo* (mini-doll), and *mame-zara* (mini-plate) all denote objects that are much smaller than normal. . . . If one explores this concept of miniaturization beyond Japanese folklore into other aspects of Japanese culture, one discovers fascinating realms—of tiny festival dolls, for example, and dwarf trees (*bonsai*). And here one gets a glimpse of what is unique in Japanese culture . . . it is not so much Japan's external conditions that drove it toward smallness as an innate propensity toward shrinking things.

While "innate propensity" may be overstating things a bit, the Japanese clearly excel at using miniaturization as a competitive tool. Sharp's development of hand-held calculators is a good example. In 1964 Sharp introduced the CS-10A desktop electronic calculator that weighed 55 pounds and cost $4,100. By using integrated circuits, Sharp reduced its Model CS-16A desktop calculator to 8.8 pounds and $1,770 in 1967. By 1972 light-emitting diode (LED) screens reduced the weight of its Model El-81 to half a pound, bringing the price down to $300. By 1980 the use of energy-efficient chips reduced the weight of the Model El-868 to a quarter of an ounce; the price had plummeted to $23. Today, Sharp's solar-powered calculators can be purchased for $4–9 at any drugstore.

By rapidly miniaturizing its calculators and constantly reducing production costs, Sharp is thus able to remain competitive in the calculator market. This is the famous "learning curve" phenomenon in which manufacturing costs decline as production volumes increase. By forcing manufacturers to reduce the size and number of parts used in new products, miniaturization is a strategic tool for quickly moving down the learning curve.

Miniaturization is also a technique for developing new product concepts. In the early 1980s, for example, Plus and Company tried selling portable copiers for use with blackboards. Images and handwriting on a blackboard could be recorded by sliding a hand-held copier over them. The product failed in the marketplace and was dropped for several years, but a group of Plus researchers persisted in developing a new pocket-sized copier for businesspeople and students. In 1984 Plus introduced the "Copy-Jack," the world's first portable copier, which was shaped like an electric shaver and weighed only 440 grams. Matsushita Electric liked the concept and asked Plus to redesign the copier to make it sleeker and less expensive. In June 1985 Matsushita introduced the redesigned Copy-Jack, which caught skeptical retailers completely by surprise. On the first day of sales, the 10,000 Copy-Jacks immediately sold out, creating an instant three-month backlog of orders. Other Japanese makers quickly jumped into the mini-copier market, setting off a wave of spin-off products. NEC, Toshiba, and other Japanese computer makers, for example, have connected portable copiers to their personal computers for use as image-scanning "mice." Unexpectedly, mini-copiers have opened the door to mini-scanners for laptop and portable computers, which, like facsimile machines, will change the way we conduct business.

For new products and services, the Japanese are always pursuing a variety of miniaturization strategies, which they have used as powerful tools for reducing costs and opening up entirely new markets.

Simplicity

If miniaturization is a source of refinement, simplicity is its cultural handmaiden. Whereas Americans and Europeans often develop complex, large-scale solutions to problems, the Japanese constantly pare down and reduce the complexity of products and ideas to the barest minimum. They streamline the design, reduce the number of parts, and simplify the inner workings and moving parts. The influence of Zen and haiku poetry are often evident in the simplicity and utility of Japanese designs. O-Young Lee observes this preference for simplicity in Japanese dry landscaping: "By paring, eliminating, cutting, peeling, and throwing away, a natural form not unlike a protoplasm results. Pare down and pare down some more. Discard the useless, the decorative. Peel away the layers that lie on nature as you would peel an onion."

Fuji Photo Film's disposable camera, dubbed "Utsurun-desu" ("It takes a picture"), is an example of how Japanese companies simplify technology to its bare essentials. In 1985 the manager of Fuji Photo Film's Consumer Photo Product Division, Keiji Nakayama, headed a team that combined optical technology with industrial design. Their goal was to develop a disposable camera for teenagers and young adults. The camera had to be easy to use and offer high-quality pictures, but they also wanted something that could be sold for less than $12 and that did not look like a high-tech, sophisticated camera, to avoid scaring away some customers. To achieve its goal, the Fuji Photo Film team pared down its camera. A paper box design was chosen to reduce costs and weight. The lens system was simplified by eliminating the focusing and diaphragm devices. As the *Japan Economic Journal* noted: "A compact, box-shaped unit was designed in order to overcome resistance to the 'use-it-only-once' idea, with emphasis on attractiveness and simplicity—no film to unwrap and put into the camera, no doors to fumble with and open—just a small, colorful 'magic' box." To appeal to the youth market, the camera was offered in four bright colors and came in a colorful protective bag.

When the Utsurun-desu was launched in July 1986, its phenomenal success even caught company president Minoru Onishi by surprise. Fuji Photo Film's stock rose sharply when the product was announced. Sales took off, outstripping the company's planned annual production of one million units. And most surprising, a customer survey revealed that the biggest users of Fuji Photo Film's disposable cameras were not teenagers but businessmen in their thirties. They wanted something lightweight and simple for their business trips.

Thus, as Fuji Photo Film learned, simplicity is an art in its own right, one that often leads to unexpected results.

Incompleteness

If simplicity focuses on reducing the complexities of life, the Japanese concept of incompleteness (*mikansei*) is its cultural opposite. Since Japan's feudal period, the notion of leaving something slightly incomplete or partially undone has been a common feature of Japanese art and etiquette. Both the artist's canvas of a large white expanse touched only with a few suggestive brush strokes and the asymmetric incompleteness of a Zen garden convey the notion of unfulfilled possibilities. They elicit creative responses by leaving viewers to "fill in the gaps" using their own imaginations.

Japanese companies use the notion of incompleteness to encourage their employees to develop new ideas. From a management perspective, a product—like a human being—is never complete but requires constant modification and polishing. Constantly refining a product is a never-ending process. Given a state of incompleteness, employees can become creative "artists" who are encouraged to re-create products and services in new ways.

The evolution of the 35mm single-lens reflex camera is an example of the incompleteness philosophy at work. During the late 1970s, sales at Japan's leading camera companies sagged as the market for 35mm cameras became saturated. Like previous lulls, many industry watchers thought there was little future in the camera business. Sony had failed with its Mavica optical disk camera and others fumbled with electronic cameras. But in the early 1980s, Nikon hit upon the point-and-shoot 35mm camera for amateurs, which triggered a rush of new family-oriented cameras. Then, Canon hit the market with its best-selling EOS automatic 35mm camera, which immediately made it the leading company. That success was followed by Minolta's stunning Maxxum 7000 automatic camera, which catalyzed major breakthroughs in electronic systems. Canon, Nikon, Minolta, Pentax, and other makers added ergonomically designed handgrips, electronic exposures, and other new features to a "mature" product.

What this rush of product creativity suggests is that Japanese companies are capable of viewing their products as "incomplete," then adding new features to transform them into hot-selling products. Incompleteness is not seen as a sign of weakness, but as a window on future opportunities.

Visualization

Depicting ideas as visual images or refining the visual appearances is another technique favored by the Japanese because of their image orientation and strong design capabilities. Whereas Western languages are phonetically and aurally based, the Japanese and Chinese languages are more visually based because of their dependence on complex ideograms to convey meanings. While Japanese characters (*kanji*) are difficult to master, the rigorous learning process teaches Japanese to recognize complex patterns and to think in visual terms. The written language is a form of visual shorthand—very much like stenography or international street signs. For example, the Japanese will often express a complex idea using a single *kanji.* The character *wa* (harmony) conveys a whole array of emotions and ideas to Chinese and Japanese that is not easily translated into Western languages. With the aid of computer-aided design (CAD) systems, the use of visual icons is spreading as the Japanese begin experimenting with visualization techniques to design new products.

Daiwa House and Kikusui Homes, for example, have developed "design-your-own-home" computer design systems that allow customers to choose the number and size of rooms, the types of building materials, and the quality of various amenities. The computer system automatically draws the floor plan layout, provides a visual image of the final product, and calculates the final closing costs, mortgage payments, and qualifying income required for the home. Bedrooms, living rooms, kitchens, and other home features are denoted by a few *kanji,* and designing one's home becomes merely a matter of moving boxes and icons around on a two-dimensional plane, which is then translated into three-dimensional drawings. By automating and visualizing the design process, buyers can choose from a variety of housing designs without having to invest a lot of time and money. Moreover, they are encouraged to visualize and experiment with their own design ideas, which makes for much happier customers.

As personal computers and CAD systems proliferate throughout Japan, new visualization techniques will become increasingly important for training people in research and product design. NEC, for example, is developing four-dimensional workstations capable of simulating three-dimensional objects over time. But it will be the ability of users to visualize the desired end result that will unleash the computing power of these new machines.

Transformation

Paring down, miniaturizing, and simplifying things are practices deeply rooted in Japanese tradition because of the premium placed on space and limited resources. Another favorite technique is transformation—rearranging parts of a setting or system to develop something entirely different. Perhaps the best example is the Japanese house, which traditionally features multi-function rooms that serve a variety of purposes. A living room can be transformed into a master bedroom by merely rearranging sliding screens (*shoji*), putting away collapsible furniture, and pulling out bedding (*futon*). Bathrooms double as laundry rooms, and dining rooms turn into family TV rooms by moving around several pieces of furniture.

Because of the high cost of land, Japanese companies have adapted this technique to shopping malls and office complexes. The Nakagin Capsule Tower Building in the Ginza district of Tokyo, for example, is a modular building that can be transformed from one shape to another over time. The building consists of 140 capsule-shaped rooms measuring 8 × 13 feet, which can be rearranged and stacked like building blocks into different shapes as needed. Given the dynamic ebb and flow of urban businesses, the concept of "transformer buildings" may replace the outdated notion of inflexible, disposable buildings that are demolished after several decades of use.

The concept of transformation is rapidly spreading throughout Japanese industry because of its commercial potential. For example, the "Transformer" toy, which can be changed from a sportscar into a robot, is being copied by Japanese carmakers. Mazda has developed the MX-04 concept car that allows the owner to change the car's body style at will. By removing the body panels and replacing them with another type, car owners may eventually be able to transform their daily runabouts into sporty weekend vehicles. Given the shortage of garage space, most Japanese can only afford to own one car, so "transformer cars" are being explored as one possible solution to rapidly changing consumer tastes.

* * *

In this chapter, we have explored some of the creative refinement techniques that Japanese companies employ to develop new, innovative products and services. For years, Westerners have considered Japanese refinement techniques to be purely imitative. Consequently, we have ignored the creativity and imaginativeness with which Japanese companies have used these techniques. The question arises: If Japanese companies have been so adept at refining products, what will happen when they pursue leading-edge technologies? Will they become technology leaders? Can Japan become a major force in scientific research? In the next section, we consider these questions as we examine several emerging technologies where Japanese companies have achieved world-class status—high-definition television, next-generation computers, and superconductivity—and analyze how Japanese companies are applying creativity techniques to these new challenges.

PART III
In Search of Creative Breakthroughs

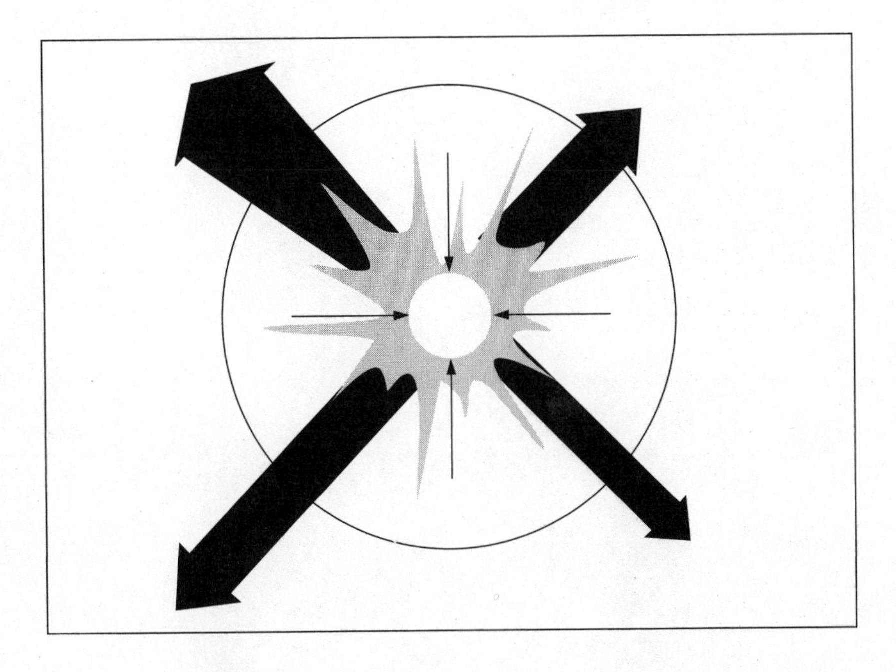

10

High-Definition Television: The Next-Generation Video Battlefield

The Japanese government has taken off the brake and is putting its foot down on the gas.

—Sosuke Yasuma
Director of HDTV Programming
Japan Broadcasting Corporation

DURING the 1988 Seoul Olympics, Japan Broadcasting Corporation (NHK) conducted a nationwide experiment that set the stage for the next revolution in video and imaging technologies. For two weeks, almost four million Japanese viewed crystal-clear broadcasts of Olympic events from 200 wide-screen "Hi-Vision" sets located in major department stores, video exhibition halls, and corporate galleries throughout the country. NHK's Hi-Vision test was a huge publicity success, comparable to the unveiling of Japan's first color TVs at the 1964 Tokyo Olympics.

In the months leading up to the Seoul Olympics, Japanese video makers had dazzled the country with a tantalizing array of state-of-the-art video cameras, 36-inch television screens, high-resolution videotape recorders (VTRs), videodisk players, and 400-inch projection screens, all of which far exceeded the performance of existing video equipment. In fact, the Olympics showing was so successful that NHK began experimental Hi-Vision broadcasting in June 1989, almost a year earlier than originally scheduled. Priced beyond the reach of the average viewer, the Hi-Vision showcase was nevertheless more than an overheated sales pitch; it was a display of raw technological prowess. "Television was initiated by the United States," proclaimed Keiji Shima, the chairman of NHK, "but Hi-Vision will be led by Japan." If the United States had its space shuttle and South Korea the 1988 Olympics, Japan wanted to show the world it was the acknowledged leader in video technology and, potentially, the leader in next-generation videocomputing.

NHK's Hi-Vision gambit is a high-stakes wager on the future. While the Europeans and Americans played it safe in Seoul with conventional TV broadcasting, NHK sent thirty-seven members of its HDTV broadcasting team, outfitted with portable cameras with twice the sensitivity of normal TV cameras and with digital editing VTRs for composing the action on the spot during Olympic events. The edited images were sent over optical fibers to a relay station, which beamed them by INTELSAT-V (International Communication Satellite V) to ground stations in Tokyo and western Japan. NHK then transmitted the pictures via Japan's BS-2b direct broadcast satellite to Hi-Vision receivers in eighty-one locations throughout Japan. For several hours each day, thousands of people crowded the Hi-Vision showrooms for a glimpse of live Olympic coverage in high-resolution detail. Unlike ordinary TV, the clarity of Hi-Vision broadcasting was stunning. Swimmers churning the waters at both ends of the Olympic pool were clearly visible at the same time. It was a total audiovisual experience, like being in Seoul itself.

The Japanese "Muse"

NHK's experiment with HDTV—also known as advanced TV (ATV) in the United States and Europe—is Japan's bold attempt to take a commanding lead in TV broadcasting and its spin-off technologies, such as VTRs, video disks, interactive video software, advanced office and home computing, semiconductors, and color graphic terminals. With this advanced system, NHK has set the agenda for the next round of video technologies and has prodded Japanese companies into pursuing more creative research. Hi-Vision is yet another reminder of Japan's uncanny ability to recycle old technologies with an injection of fresh new ideas.

Since RCA's introduction of black-and-white TVs in 1940, television has undergone dynamic change. Color television, Sony's Trinitron gun technology, cable TV, satellite broadcasting, large screens, and VTRs have dramatically reshaped television viewing habits during the last twenty-five years. Today, picture reception is better, TV screens are brighter, and programming is more varied than ever. But despite these improvements, current TV broadcasting still relies on old technology developed before World War II. In the United States and Japan, televisions use RCA's original system of 525 scanning lines per screen (the more horizontal scanning lines, the sharper the image), which was standardized by the U.S. National Television System Committee (NTSC). For small TVs, the 525-line system is fine, but on large screens over forty inches the fuzzy edges and dull, mushy colors are unappealing to the average viewer. In Europe, viewers benefit from much sharper images broadcast from the 625-line German PAL (phase alteration line) system and the 825-line French SECAM (sequential with memory) system. Although the Japanese have developed an improved NTSC system with 525 lines, this system has become obsolete with the introduction of large-screen TVs and projection TVs. While the U.S. broadcasting industry ponders how to adopt a new system without making existing TV sets obsolete, Japan has decided to go ahead on its own.

Japan's push into HDTV technology came in 1968 during the Mexico Olympics. NHK saw the limitations of the 525-line NTSC system and began studying ways of developing a better system. Video engineers felt that television should not just transmit clear images but should also convey a feeling of dynamism and "you-are-there" presence. For example, existing TVs cannot show in sharp resolution an entire football field from goalpost to goalpost. The overall picture is generally fuzzy, and the edges are distorted. Moreover, for sophisticated programming, such as rendering the subtle colors and shadings of paintings, sculptures, and other works of art, conventional TV is totally inadequate. To determine the features of the ideal TV of the future, NHK video engineers conducted in-depth surveys to analyze viewer preferences, comfortable viewing distances, differing screen sizes, and other factors.

They found that the preferred viewing distance is two to four times the height of the screen, suggesting a 1,750-line system. But the bandwidth required was impractical for commercial broadcasting because it threatened to crowd out other signals. So NHK settled on a 1,125-line system that transmits thirty frames per second on a large screen with a five-to-three aspect (width-to-height) ratio—matching the angle of the field of human vision. To accommodate the 1,125 scanning lines, the video signal bandwidth was increased from 4.2 megahertz to 20 megahertz.

With this system, NHK is able to transmit TV pictures with a quality equal to that of 35mm or 70mm big-screen movies. For example, NHK's Hi-Vision sets can display an entire baseball field showing the pitcher, batter, runners, and outfielders simultaneously, as though viewed from an infield seat. Although fewer close-ups would be shown, broadcasting studios could zoom in on a scene or TV viewers could point laser guns to portions of the screen to create enlarged video "windows" without losing clarity or resolution. These enlargements could be either recorded in real time or "frozen"; then they could be stored in VTRs for later editing or printed out in color on laser printers. This feature would be the first step toward home video editing and, eventually, home videocomputing.

But to develop this system, NHK had to come up with several technologies, including entirely new cameras, TV receivers, and broadcast transmitters. One development was a high-resolution camera tube, using a diode-gun impregnated-cathode Saticon (DIS) for 1-inch and ⅔-inch portable and studio cameras, which have a horizontal resolution of 1,200 lines or more. These are now being used in commercial Hi-Vision video cameras. Another innovation was large 40- to 100-inch color terminals for Hi-Vision TVs, VCRs, video copiers, and video disk players. Collaborating with leading Japanese video makers,

NHK developed a 40-inch direct-viewing picture tube using reinforced glass walls to withstand the increased atmospheric pressure. Since larger screens using traditional glass tubes cannot feasibly be made, NHK is working on flat TV screens and 100-inch reverse-projection displays using translucent screens for hanging on the wall. For practical wall-hanging TV screens, Japanese companies are working on plasma panel displays that use arrays of small light-emitting cells. These display technologies are also being used in desktop and laptop computers and large graphic terminals.

At the same time, NHK pursued several other innovations to handle the wider video bandwidth required for its Hi-Vision system: digital circuits, video bandwidth compression, and direct satellite broadcasting. These technologies will have broad applications in the future.

Digital circuits are essential for developing crystal-clear Hi-Vision systems because they provide higher video and sound quality. Japanese video makers now design advanced semiconductor devices in their TVs, VTRs, and video disks to reduce the size and cost of Hi-Vision systems. With their lead in memory chip technology, companies such as Fujitsu, Hitachi, Matsushita, NEC, and Sony have developed large-capacity video memory chips to handle special effects. Hi-Vision TVs, for example, feature from eleven to thirty-one megabits of memory chips to produce special effects, such as zoom, split screen, freeze-frame, instant-replay, image storage, and ghost removal. Japanese companies have also developed specialized chips to process digital signals. In 1988, for example, Hitachi and Mitsubishi jointly developed a set of compact chips capable of processing Hi-Vision signals. And, of course, CCD (charge-coupled device) sensors capable of picking up faint images in dim lighting have been used for years in hand-held video cameras. But according to Dr. Motokazu Uchida, general manager for research and development at Hitachi, "There is still a lot of work to be done on both high-speed signal processors and the video RAM chips needed for low-cost manufacturing of HDTV sets to become practical."

The biggest problem with NHK's original Hi-Vision system

is that it requires twenty megahertz of video bandwidth—or over three times the six megahertz used for current TV broadcasting—making it unfeasible for over-the-air commercial broadcasting. NHK tackled the problem and, in January 1984, announced a technical breakthrough that reduced the Hi-Vision signal bandwidth. Called the MUSE (multiple sub-Nyquist sampling encoding) system, this bandwidth compression technique involves squeezing 20-megahertz Hi-Vision signals into a single 8.1-megahertz channel, thus requiring much less bandwidth. MUSE divides the broadcast image into still and moving portions, then recombines them using frame memories. With this method, TV stations can transmit images two to three times clearer than existing 525-line NTSC sets, using only slightly more bandwidth. Suddenly, Hi-Vision became a commercially viable technology.

But the MUSE system still requires satellite broadcasting because its bandwidth-hungry signal cannot be easily accommodated in the crowded over-the-air radio spectrum. To overcome this barrier, Japan is developing a direct broadcast satellite (DBS) program with the National Space Development Agency (NASDA). In April 1978 NHK put an experimental broadcast satellite into orbit to test the picture and sound quality of Hi-Vision signal transmissions. In January 1984 Japan launched its first DBS, BS-2a (also known as YURI 2a). Unfortunately, two of the three transponders (radar sets that emit radio signals of their own in response to receiving a signal) failed, setting back NHK's Hi-Vision experiment, as well as domestic sales of small dish antennas. In February 1986 a three-channel back-satellite, BS-2b (YURI 2b), was launched, but its altitude control system failed. NHK General and NHK Educational Television used the remaining two channels to begin broadcasting in December 1986.

Since 1985, NHK has produced several Hi-Vision programs—including "The Yellow River," coproduced with China—that have been converted to the NTSC system and broadcast by BS-2a. In July 1987 NHK General and NHK Educational Television combined their efforts and began twenty-four-hour DBS programming. Half of the day is devoted to

"World News," which carries national and international news and stock market information from the New York, London, and Tokyo exchanges. "Super Stadium" provides coverage of international sporting events, foreign films, theater and opera performances, and philharmonic concerts. Combined with pulse-code modulation (PCM) audio broadcasting technology, DBS programming offers extremely high-resolution TV viewing with clear digital sound. By bouncing its signals off the BS-2b satellite located in a stationary orbit 36,000 kilometers over the equator, NHK is able to provide DBS programming to remote mountainous areas and islands. DBS home satellite antennas originally cost over $1,300, but prices are rapidly coming down as the competition heats up. Yagi Antenna, for example, sells a small 48-centimeter–diameter dish for $410, while Toshiba's large 75-centimeter–diameter antenna runs $920, including installation. As of mid-1987, over 143,000 households were receiving DBS service, including 100,000 in poor reception areas.

Full-scale satellite broadcasting will begin in 1990 when NASDA launches the three-channel BS-3a satellite. Two NHK channels and one from Japan Satellite Broadcasting will provide one to two hours of daily programming, plus facsimile and data broadcasting. By 1990 NHK expects to sign up about 1 million households. In 1991 the BS-3b will be launched, offering one channel for Hi-Vision broadcasting. Both geostationary satellites will transmit 12- to 14-Gigahertz signals to 75-centimeter–diameter home satellite dishes. The launch effort will be a joint undertaking: the H-I and H-II launch vehicles will be made by Mitsubishi Heavy Industries, the satellites by RCA, and the traveling wave tubes by NEC. By 1992 NHK's total viewership is expected to reach 3 million of the 32 million households in Japan, or 10 percent.

In 1997 NHK plans full-scale Hi-Vision broadcasting when the eight-channel BS-4 satellite is launched. The Ministry of Posts and Telecommunications (MPT) may allocate four of the eight channels to Hi-Vision broadcasting if there are enough home satellite antennas to warrant it, leading to a Catch-22 dilemma: if not enough channels are allocated to Hi-Vision, home-owners may not buy enough antennas to justify four Hi-Vision channels. Private Japanese satellite companies may decide to form their own consortium to launch a Hi-Vision broadcast satellite if the government hesitates. By then, a wide variety of Hi-Vision applications may require more than one satellite.

Japan's Hi-Vision Strategy

The Japanese government is bullish about Hi-Vision television because of its potential for expanding the domestic economy by creating lucrative new markets and giving Japanese companies an edge over foreign competitors. Both MITI and MPT have formulated their own industrial policies to jump-start the infant Hi-Vision industry into action. These policies are similar to those used in the past to target makers of memory chips, supercomputers, optoelectronics, robotics, and fifth-generation computers. They involve organizing joint R&D projects and industry forums, providing tax incentives and R&D subsidies, and setting technical standards.

In September 1987 MPT issued its forecast of Japan's Hi-Vision market, assuming that the BS-3a, BS-3b, and BS-4 satellites would be successfully launched. MPT forecasts a Hi-Vision broadcasting market of $26 billion in the year 2000, with cumulative sales exceeding $110 billion (see Table 10–1). Film production, video packages, video theaters, printing, and publishing will be much smaller, more immediate markets. These forecasts do not even consider the more profitable computer sector, which will merge with HDTV in several years.

For the time being, however, Hi-Vision equipment makers are faced with the prospect of limited markets because of high initial production costs. To overcome this obstacle, MPT is promoting HDTV standards, broadcasting, and public awareness. In June 1987 MPT issued a report that called for financing 50 percent of Hi-Vision development costs and an annual $375 million budget. This program is part of MPT's twenty-year program to promote domestic demand for Hi-Vision television.

TABLE 10–1
MPT's Hi-Vision Market Forecast ($ billions)

Market	Market Size Year 2000	Cumulative Sales to 2000
Broadcasting	$25.64	$109.12
Film production	.39	2.12
Video packages	.11	.33
Theaters	.018	.16
Printing and publishing	.012	.12
Total	$26.17	$111.85

NOTE: $1 = 130 yen
SOURCE: Ministry of Posts and Telecommunications.

Most of this research will be privately financed. To help boost sales of Hi-Vision equipment, MPT has established four Hi-Vision promotion programs. The first was the nationwide Hi-Vision demonstration project during the 1988 Seoul Olympics. The second is a private Hi-Vision investment fund of $100 million to finance Hi-Vision software production, video libraries, transmission and receiver systems, and equipment leasing. The third is an extensive research program to develop terrestrial Hi-Vision broadcasting, system compatibility with simpler extended-definition TV (EDTV), small dish antenna development, home receiver usage, display technologies for large-screen home viewing, and compatibility with the existing 525-line NTSC system. The fourth is a plan to establish ten experimental Hi-Vision model cities nationwide to test the feasibility of next-generation HDTV systems for international conferences, resorts, health and welfare, cultural activities, high-tech businesses, entertainment, and university classes. In October 1988 twenty-three candidate regions submitted their Hi-Vision model city concept plans; eighteen cities were chosen.

MITI, which is responsible for promoting industrial and public uses, is more bullish about the Hi-Vision market than MPT. MITI forecasts Hi-Vision sales exceeding $32 billion by the year 2000 (see Table 10–2). Two-thirds of the revenues will come from equipment hardware, and one-third from video software production and use. Both ministries believe that broadcasting will be the major market in the year 2000, but MITI is

TABLE 10–2
MITI's Hi-Vision Market Forecast
($ billions)

Market	1995	2000
Hi-Vision equipment	$6.2	$20.8
Consumer use	5.4	13.1
Industrial use	0.8	7.7
Image software production	0.3	3.8
Image software use	1.4	7.7
Total	$8.9	$32.3

NOTE: $1 = 130 yen
SOURCE: Ministry of International Trade and Industry.

more optimistic about industrial applications, such as factory monitors and simulation equipment. MITI estimates that 10 percent of Japan's 35 million households will have Hi-Vision by 1995, and that this figure will double by 1998 after Japan launches its BS-4 satellite.

In 1988, to assist the industry, MITI established Hi-Vision centers in its "New Media Community Cities" to promote next-generation media technologies. In these cities, MITI is promoting the use of Hi-Vision systems in schools, museums, libraries, hospitals, publishing, manufacturing, distribution, and real estate (see Table 10–3). One popular application is "hypermedia," in which images, videotapes, narration, and text can be quickly retrieved for educational and industrial purposes.

For both ministries, Hi-Vision is viewed as an extremely attractive economic growth generator. Besides boosting sales in the stagnant color TV and VCR industries, it will improve Japan's competitiveness against Asian countries like South Korea and Taiwan. It will also be Japan's stepping-stone into high-end industrial markets, such as electronic publishing and advanced image computing. Japanese video makers have been desperately seeking "the next VCR" and believe they have found it in Hi-Vision.

Promoting Practical Applications

Hi-Vision systems have many potential applications, but finding applications that can be immediately commercialized is another

TABLE 10–3
Public and Industrial Hi-Vision Applications

Field	Hi-Vision Application
Libraries	Science and technology image information retrieval for scientists and nonprofessionals Text retrieval system using optical disks
Hospitals	X-ray diagnostic photo data base Home care video examinations
Museums	Painting, sculpture, and archaeological image data bases Background information for art exhibitions Intermuseum image information sharing
Schools	Classroom teaching video network Video materials for science and social studies Anatomical images and medical techniques
City services	Town information centers Introductions to regional visitor spots
Publishing	Video image hard copy printing Simultaneous text and image printing Continuous-operation recording and printing Simultaneous display of publication pictures
Video production	Shift from film editing to video editing Improved video editing quality
Movies	Digital image processing for movie-making Videocassette production Automated television camera modulation Historical film videocassette libraries Mini-video theaters
Distribution	Department-store events and information Advertising production
Other uses	Conferences and seminar broadcasting Television conference image system Real estate information system

SOURCE: Ministry of International Trade and Industry.

matter. Although Japanese video makers have developed advanced Hi-Vision equipment, they are faced with the problem facing nearly all pioneers in innovative technology: the lack of useful applications and software. As the *Nikkei Electronics* magazine noted after the 1988 Olympics:

> Hi-Vision equipment is already being used for industrial applications. But Hi-Vision managers are won-

dering whether it will ever become a business. Hi-Vision equipment is expensive, so video software is not being developed and the market is not growing. The market is not growing, so equipment is not getting much cheaper. It is a real dilemma.

To overcome this "application gap," video makers and the Japanese government have sponsored a variety of promotional campaigns (see Table 10–4). In 1985 NHK displayed its MUSE Hi-Vision system for six months to over 20 million visitors to the Tsukuba Expo '85. In early 1987 NHK set up a Hi-Vision display in Washington, D.C., to influence policymakers responsible for setting standards. Beginning in 1987, NHK began an intensive campaign in collaboration with industry to sponsor Hi-Vision study groups, international colloquia, test broadcasting, and Hi-Vision fairs. NHK, MPT, broadcasting companies, and home electric appliance makers formed the Hi-Vision Promotion Council to educate policymakers and raise public awareness about the merits of Hi-Vision technology. In 1987 MITI established a HDTV Promotion Center to promote equipment standards and leasing, software production, and international cooperation. The campaign reached a crescendo in July 1988 when NHK transmitted Hi-Vision programs between

TABLE 10–4
Major Japanese Hi-Vision Promotional Events

Date	Event
April 1985	Hi-Vision displayed at Tsukuba Expo '85
January 1987	NHK displays Hi-Vision in Washington, D.C.
March 1987	Hi-Vision Promotion Group (6 months)
June 1987	Hi-Vision Fair
October 1987	Canada HDTV Colloqium; Telecom '87
October 1987	NHK and KDD test Intelsat-V Hi-Vision
June 1988	Hi-Vision Fair
July 1988	Hi-Vision display between Nara's Silk Road Expo and Australia's International Leisure Expo
September 1988	1988 Seoul Olympics
1990	Asian Olympics; World Cup soccer

SOURCE: Ministry of Posts and Telecommunications and Ministry of International Trade and Industry.

Nara's Silk Road Expo and Australia's International Leisure Expo. Fair-goers in both countries were able to see each other's exhibitions live via three 135-inch projection receivers and eighteen 40-inch Hi-Vision sets displayed at the Japan Pavilion in Brisbane. The signals were relayed by three geosynchronous satellites: Japan's CS-2b communication satellite, Intelsat-V, and Australia's AUSSAT.

NHK's publicity blitz peaked with its Hi-Vision broadcast of the 1988 Seoul Olympics, which was a classic example of how Japanese industry works closely with government ministries to develop new markets for an emerging technology. As noted by Yoshinobu Numano, chief director of NHK's Satellite Broadcasting Headquarters, the Seoul Olympics marked the end of Hi-Vision's experimental phase and the beginning of its commercial use. In June 1989 MITI and forty-nine Japanese companies established Hi-Vision Communications to plan HDTV events, produce software, and lease equipment.

Clear Vision—Competing with NHK's Hi-Vision

NHK's Hi-Vision publicity campaign has not gone unchallenged by video makers and rival Japanese broadcasting companies, which have teamed up to introduce two competing systems: improved-definition TV (IDTV) and extended-definition TV (EDTV). Why are these systems being developed? And how do they differ from NHK's Hi-Vision? The main reason has to do with cost and system compatibility. NHK's Hi-Vision offers very high resolution, but at a premium because of its incompatibility with current NTSC standards. Hi-Vision viewers must buy very expensive TV sets and satellite dish antennas. During the 1988 Olympics, Hi-Vision sets were selling for $80,000 and satellite antennas for $2,000, putting Hi-Vision out of the reach of most households. Meanwhile, IDTV and EDTV sets were selling for under $2,000. Even if Hi-Vision TV prices decline to $3,500 in the mid-1990s as expected, their usage will be limited to businesses in highly competitive markets, such as bars, department stores, and restaurants, where Hi-Vision sets can be used for promotion. Regular Hi-Vision broadcasting will not begin until the launching of BS-3a in 1990, and widespread use is not expected until the eight-channel BS-4 satellite is launched in 1997. As a result, Japanese broadcasters and video makers view IDTV and EDTV as stepping-stones to full-scale Hi-Vision.

IDTV is compatible with the existing 525-line NTSC standard and has been available to the Japanese public since 1986. To achieve its clarity, IDTV uses digital processing techniques and five to seven megabits of video memory chips to eliminate snowflakes and ghosting. Hitachi, Matsushita, Sony, Toshiba, and other Japanese video makers already sell IDTV sets for about $2,000–2,500—twice the price of existing sets due to the high cost of video memory chips. But IDTV prices will decline as video competition intensifies and memory chips become cheaper.

Meanwhile, private broadcasters belonging to the Japan Television and Broadcasting Association (JTBA) hope to grab market share from NHK by developing their own EDTV system, dubbed "Clear Vision." Compatible with existing TV signals, EDTV sets can also receive specially encoded EDTV broadcasts. The system is being developed in two phases. Version I does not require satellite transmission and can be viewed on 525-line TV sets without special adapters. Version II will require wider TV screens, but the standards have not been set. Advanced digital signal processing and megabit memory chips will increase the horizontal scanning by 40 to 50 percent, making for very clear images. Although Clear Vision broadcasters have to upgrade their cameras and transmission equipment to handle ghost-cancellation signals, most improvements can be made in the TV sets. JTBA has already tested EDTV broadcasts, and private companies began regional EDTV broadcasts in September 1988. Matsushita and other video makers have introduced 30-inch EDTV sets priced at $2,300, double the price of current 30-inch sets. Beginning in March 1989, full Clear Vision broadcasting will be available from five Tokyo stations and four Osaka stations. Thus, the battle among IDTV, EDTV, and NHK's Hi-Vision is rapidly heating up.

The Global Battle Over Standards

The battle for control over global high-definition standards is also intensifying. NHK has a five-year technological lead with its MUSE system, but strong resistance from Europe and the United States, which fear a replay of Japan's takeover of consumer electronics, has prevented NHK from establishing MUSE as an international standard. Instead, competing HDTV standards are emerging in Europe and the United States.

In Europe, the British Broadcasting Corporation (BBC) has proposed a system using 1,501 lines per frame and thirty megahertz of bandwidth, while Philips is a proponent of its HD-MAC system. In June 1986 Robert Bosch, Philips NV, Thomson, Thorn/EMI, and over ninety other European electronics companies formed EUREKA 95 to develop a European HDTV standard by 1990 that will feature 1,250 scanning lines. An intermediate standard features 1,152 lines.

In the United States, HDTV systems are proliferating like alphabet soup. The David Sarnoff Research Center in Princeton, New Jersey, has unveiled a single-channel ACTV-I (advanced-compatible) high-definition TV system and a second-stage, two-channel ACTV-II. North American Philips, a subsidiary of the Dutch giant Philips NV, has introduced HDS-NA (High-Definition System for North America). William E. Glenn, former research director of CBS Laboratories and now at the New York Institute of Technology (NYIT), has developed the NYIT system using one extra channel and a compatible signal encoding method. The small Del Rey Group has proposed a compatible system that would cram an HDTV signal into a 6-megahertz bandwidth.

These competing standards, however, have divided the already fractious U.S. electronics industry into warring groups; by contrast, the Japanese and Europeans present a more unified front. To prevent the Japanese and Europeans from totally dominating the U.S. HDTV market, which could reach $145 billion over the next twenty years, U.S. electronics industry leaders successfully lobbied the Federal Communications Commission (FCC) to set independent U.S. standards. In September 1988 the FCC issued technical guidelines known as Advanced Television (ATV) Document #MM88–268, which requires that the U.S. ATV system be compatible with conventional television receivers. The ruling encourages a gradual changeover to HDTV in the early 1990s, similar to the way the industry shifted from black-and-white to color in the past. It effectively removed NHK's Hi-Vision as a candidate system. The FCC's next step is to establish full technical standards, a job that will be complicated by the more than twenty proposals already submitted for consideration.

But even with independent U.S. standards, it is highly unlikely that American companies will each invest the hundreds of millions of dollars needed to re-enter the competitive HDTV market. Wall Street looks for quicker returns on investment, especially in areas where the Japanese are major competitors. Moreover, Katsumi Osuga, director of HDTV promotion at MPT, believes Japan's HDTV vendors could easily satisfy new FCC standards. The American National Standards Institute (ANSI) recently approved NHK's Hi-Vision system as a studio standard for the ATV system in the United States, which gave a boost to Sony and other Japanese HDTV makers.

Faced with this dilemma, the U.S. electronics industry is trying to rally its forces. In December 1988 the American Electronics Association (AEA) opened the HDTV Information Center in Washington D.C., to lobby for new regulations and legislation that would help the U.S. HDTV industry. In May 1988 it called for the U.S. government to spend $1.35 billion on advanced video research. The Department of Defense also announced plans to fund a $50 million HDTV research project to develop new compact display technologies that it will need for combat helicopters, training simulators, and other defense systems. Besides preventing total domination by European and Japanese companies, the Pentagon views HDTV as a way to maintain a domestic ability to manufacture semiconductors. The HDTV consortium will complement the Semiconductor Manufacturing Technology (SEMATECH) consortium, which is developing next-generation chip technology.

As a result of these European and U.S. countermeasures, NHK, MPT, and Japanese TV makers have virtually given up trying to establish the MUSE system as an international standard. Instead, MPT began transmissions tests of its cable TV networks in the Tokyo region in October 1988, and NHK began experimental satellite Hi-Vision broadcasting in June 1989, with full-scale broadcasting scheduled for 1990. At the same time, Europe is planning to have experimental HDTV broadcasting in place, using its HD-MAC system, in time for the 1992 Olympics and liberalization of the European Common Market.

Is It a Television or a Computer?

The battle over high-definition TV is not just a struggle for control over next-generation television. It is a prelude to the much larger confrontation looming in the early 1990s: the battle for dominance in next-generation video- or multi-media computing. Richard J. Elkus, Jr., chairman of Prometrix Corporation and a leading HDTV proponent, argues:

> The HDTV market represents a complete merging of technologies essential to the domination of the consumer electronics market and major portions of the telecommunications and computer markets. The nation which becomes the key supplier of production, distribution, and receiving equipment within the HDTV markets will have developed a principal means to control a major portion of the semiconductor chip and equipment business and, therefore, have achieved a significant step toward total technological superiority in the years to come.

Despite renewed interest by U.S. executives, Cees Koot, chief executive and managing director of semiconductor operations at Philips NV, is skeptical that the United States can catch up with Japan and Europe in HDTV. He told a *San Jose Mercury News* interviewer:

> I went to a trade show in Japan recently and I was completely flabbergasted by what I saw. For instance, they were showing 14-inch liquid crystal display televisions. You can't imagine how thin those televisions were. And there was an ultralight camcorder with an LCD TV in it. Very complex. I don't think American industry is able to make these. . . . What if that technology becomes a determining factor for making new laptop computers or other portable equipment? Where is America going to get that technology from?

On both sides of the Pacific Ocean, the battle lines are forming. In Silicon Valley, Steven Jobs of NeXT Computer has created a revolutionary new machine that can handle complex image data, record and play back music and voices, and store massive amounts of information on optical disks. In Japan, NHK and Sony have created Hi-Vision televisions capable of manipulating complex video images, playing back digitally recorded music and sound tracks, and storing massive amounts of video information in VTRs and video disk players. The only difference between the U.S. and Japanese machines is applications. The NeXT computer is primarily targeted at university research and the office environment, while Sony's machine is aimed at home TV viewers. But this distinction will disappear as NeXT computers move into the home office and Sony TV terminals end up in business and home office computers. Indeed, Dr. Harry Taxin, president of Sony's New Technology Center in Silicon Valley, observes: "The line between video equipment and computers is blurring. With the advent of HDTV, that line may disappear altogether." Already, Sony uses its NEEDS technical workstation in video production, and makes high-resolution graphic terminals for Apple and other U.S. computer makers. It is only a matter of time before Sony ends up in the same markets as NeXT Computer and its videocomputer clones. Indeed, their research centers are located several miles away from each other in Palo Alto, the heart of Silicon Valley.

This merging of computer and television technologies will profoundly influence the way we view the world. TV and VTR technologies are being combined into combo TV/VTR sets in Japan. During the 1990s personal computers will merge with TV sets as video software programs using hypermedia become available. Most homes in highly industrialized nations already own TV sets and VTRs, and TV sales in developing nations like China, India, and Brazil are rapidly accelerating. (Indeed, China is even exploring ways to link computers with TV sets in order to educate hundreds of millions of people via satellite broadcasting.) Most households, however, have not purchased computers because of their cost, lack of user-friendliness, and limited usefulness. Unlike televisions, computers are not very appealing, nor do they offer entertainment or learning that is not already available in conventional television. For the most part, computers still represent an intrusion of the office into the home.

The shift to high-definition television is likely to change all this as HDTV image resolution and digital sound quality improve. HDTV will be seen as not just a high-priced gimmick (as VTRs were in the mid-1970s), but as a window presenting a new way of viewing the world. With the advent of hypermedia software and other powerful video software, HDTV will become a magic carpet for students, TV viewers, classical music listeners, doctors, retailers, office workers, and artists seeking clearer images and sounds. As shown in Figure 10–1, Dentsu—Japan's leading advertising agency—sees HDTV technology diffusing into consumer, retail, medical, scientific, and telecommunications fields over the next ten to fifteen years. The early costs will be high, but they should decline far enough to make HDTV available to mass audiences by the turn of the century.

At that point, the only barrier to widespread usage will be creative vision—the ability to find new uses for these powerful new video tools. The Japanese will be major players in the development of this emerging technology. As they have done so many times before with other technologies, their creative energies are sure to transform HDTV into a myriad of new technologies.

FIGURE 10–1
The HDTV Mandala

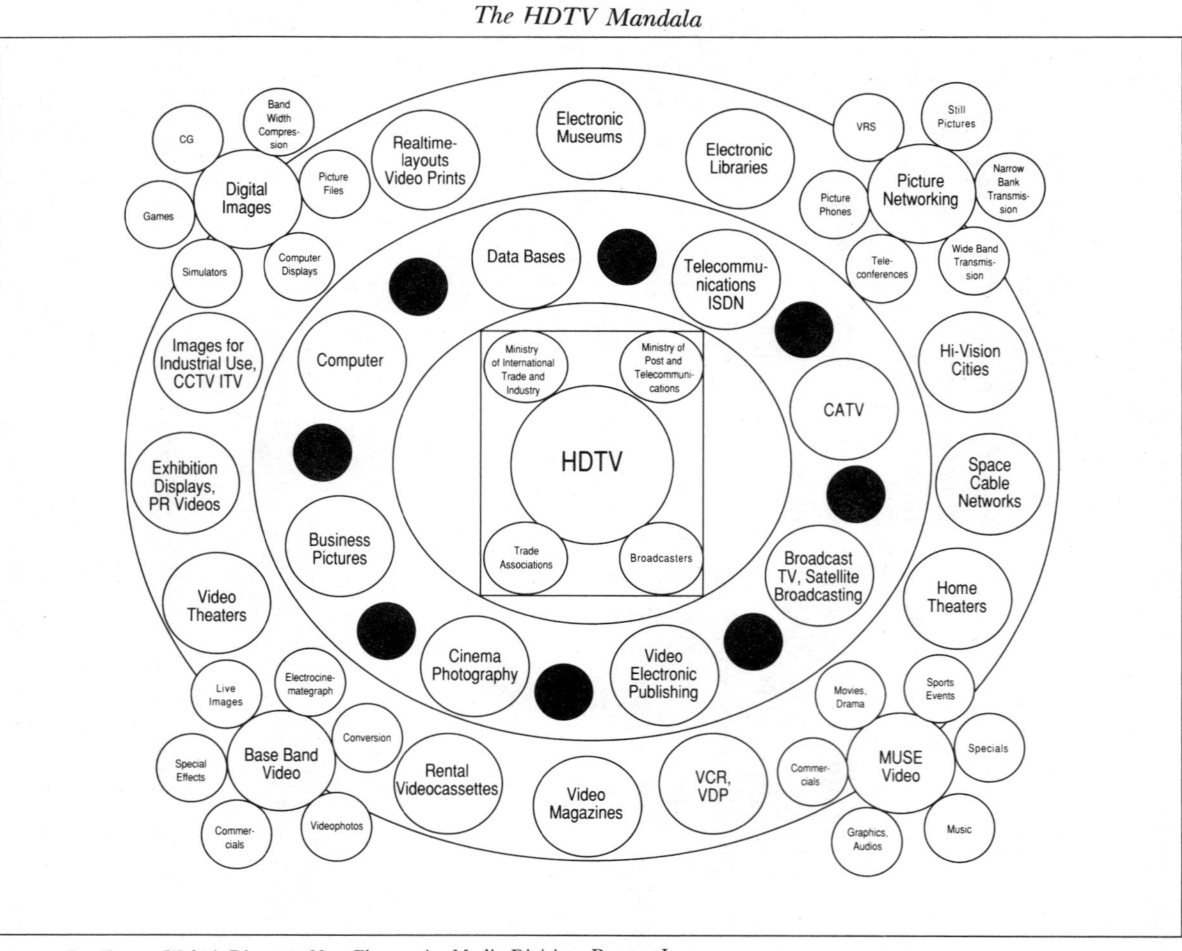

SOURCE: Dr. Kotaro Wakui, Director, New Electronics Media Division, Dentsu, Inc.

11

The Computer Bazaar of the Future

> If we are to prepare for future compatibility with applications in the 1990s, we must establish new, clean computer system architectures by doing away with outdated designs.
>
> —Ken Sakamura
> Tokyo University professor and founder of the TRON Project

WELCOME to Akihabara, Tokyo's sprawling electronics bazaar. It is April 2003. Japan's top electronics makers have just released their spring selection of new pocket appliances, which are overflowing Akihabara's crowded alleyways and videoware department stores. Displayed on shop walls are hundreds of the latest pocket computers from Casio, Citizen, Seiko, Sharp, and Sony—small, colorful plastic cards useful for figuring home budgets, learning foreign languages, building one's vocabulary, remembering faces at an important reception, monitoring one's pulse and biofeedback levels, or recording songs from the radio.

These PC cards, as they are called, feature color LCD TV screens and are operated by voice, fingerprint, or a simple push of the button. Some cards feature tiny sensors that can recognize faces, images, shapes, and movement. Others boast over sixteen Gigabits of memory space—both the Bible and an encyclopedia can be stored in them, with plenty of room to spare—yet they cost only $15. PC cards were first developed in the mid-1990s when the market for pocket calculators and laptop computers became glutted by clones from Asia and the Eastern bloc. Today they are as common as calculators.

In glass counters throughout Akihabara are fifth-generation models of Sony's "Studioman," a hand-held, three-dimensional color TV capable of simultaneously comparing and editing clips from up to five previous Super Bowls or Cannes Film Festivals, or from the latest archaeological digs in China. These hypermedia computers have enormous relational data bases that are accessible using expert reasoning systems stored in small erasable video disks, which sell for only $20. Professionals often use them for running "what if" business scenarios and complex design simulations, while students find them handy for retrieving video archives of the Louvre or Kyoto's ancient temples for class assignments. The video images, text, and animated narration are stored in ultramedia libraries developed by Hollywood scriptwriters working with university professors. Multimedia was originally invented in the United States for desktop computers, but was superceded by "hypermedia" when the Japanese developed creative new applications that made the videocomputer affordable to hundreds of millions of users worldwide.

The highlight of our tour of Akihabara is NTT's Audio Bazaar, located in a plastic geodesic dome shaped like an inverted satellite antenna. NTT Corporation has invested trillions of yen in new telecommunications and computer research, which is paying off handsomely. Inside are NTT's latest phones that can simultaneously translate international calls from Japanese to ten different languages, or vice versa, saving major corporations billions of dollars annually. These phones were better known as "fifth-generation computers" during the 1980s. NTT has won international awards for its best-selling "Silver Phones"—wristwatch videophones for senior citizens that are designed with biofeedback monitors and automatic emergency call alarms. These phones have saved thousands of lives each year. NTT also cannot keep up with the demand for its new "Page Bell" series of cellular telephone pens, lapels, and watches, which have eliminated the annoying problem of office phone tag, especially for international calls. But perhaps the most dramatic breakthrough is NTT's new holographic video-conferencing system, which transforms a normal conference room into a three-dimensional video phone call. Instead of speaking to a flat screen, the "Holophone" creates a conversation-in-the-round setting by projecting into the same room realistic color images of all attendees on both ends of the call. The first system being installed is a hot line between the White House and the Kremlin. . . .

Fantastic and farfetched? Perhaps, but these scenes of the future are not very far away. In fact, they are right around the corner. Having caught up with the West in mainframe and desktop computing, the Japanese are now in the process of creating entirely new computers that will dramatically alter the shape of the industrial and consumer world. These machines will not be pale imitations of Western computers, but rather stylish appliances and video systems that will fit right into the home and office, like a microwave oven or a high-resolution television. Unlike present computers, these systems will not require extensive computer training or technical know-how, but will be user-friendly and appealing to the average person.

What advances are the Japanese making in next-generation computing? How have they organized their advanced computer projects? And what creative breakthroughs might we expect from Japan in the future? Let's begin with what the immediate future holds—pocket computers, 32-bit computers, and supercomputers—and work our way up to the advanced systems that will emerge in the twenty-first century—fifth-generation computers, optocomputers, neural computers, and biocomputers.

From Hand-Held Calculators to Smart Cards

In late 1988 Sharp ran an alluring ad in major U.S. publications that heralded the arrival of Japanese pocket computers. "Flash Gordon said it would happen," ran the ad.

Sharp has turned science fiction into science fact with

> a remarkable new invention—the Wizard. It's the first pocket computer that helps you organize your life as never before. With incredible ease, Wizard remembers and reminds. Translates and defines. Stores memos, secrets, and world times. Even swaps data with your personal computer.

Although a novelty to Americans at the time, Sharp's pocket computer was old news to the Japanese. Two years earlier, Sharp had introduced the Wizard, then called the PA-7000, to the Japanese public. At first glance, it was indistinguishable from the hundreds of calculators vying for attention on the shop walls crowding Akihabara. But this hand-held calculator was different. Designed as a first-generation pocket computer, the PA-7000 offered a variety of functions for busy people—memo pad, personal telephone directory, address book, monthly calendar, weekly schedule, slot for program cards, and a keyboard for tapping in Japanese characters, Roman letters, numbers, and instructions. Two slot-in cards using large memory chips powered the machine: a Japanese dictionary card, and a 15,800-word English–Japanese translation dictionary card that provided the correct terms used in overseas hotels, airports, hospitals, taxis, and restaurants. Although more powerful than needed for the average person, Sharp's PA-7000 was a best-seller among Japanese businesspeople and professionals. After two years of successful sales in Japan, Sharp decided to move into the U.S. market.

Sharp's hand-held computer may not be impressive when compared with powerful 32-bit desktop computers, but it symbolizes Japan's plunge into the world of miniature computing—a bastion of Japanese creativity. Just as Japanese companies transformed desktop radios into transistor radios during the 1950s, calculating machines into pocket calculators in the 1960s, and color televisions into video Walkmans in the 1980s, they are now miniaturizing desktop computers into laptop, hand-held, pocket, and wristwatch computers. If Westerners invented computers for office users and hobbyists, Japan's creative juices are transforming them into household appliances, video toys, and everyday tools. The IBM P/S computer and Apple MacIntosh will become the "Computeman" or "Thinkman" of the future—brought to you by Sony, Fujitsu, NEC, and Toshiba.

Sharp's Wizard is the first step toward the realization of a Japanese manufacturer's dream—pocket computers. If existing computers could be shrunken, simplified, and pared down into tiny plastic cards, they could be sold to anyone, regardless of technical background and training. Computers would no longer be inaccessible office machines or yuppie accessories, but everyday tools like calculators and credit cards. And despite their technical genius, Europeans and Americans are unlikely to be major competitors because they lag behind Japan in the mass production and plastic packaging techniques needed to drive this explosive marketplace. Indeed, Japanese calculator makers, who have dominated the market since the "calculator wars" of the late 1960s, are already leading innovators in "smart card" computing.

The French invented the smart card—plastic cards embedded with microcontrollers and memory chips that enable them to compute and process information. In 1982 the French Ministry of Public Posts and Telecommunications (PPT) installed check-clearing and communication lines in three cities to reduce check-processing costs. In 1983 PPT introduced 200 public smart card phones to lessen rampant cashbox vandalism. Paymatec and Flonic-Schlumberger provided the cards. By 1985, there were more than 10,000 public phones and two million cards, driven by the French government's "Telematique Plan" involving videotext, telephones, and smart card systems. In the United States, MasterCard International and Visa began issuing smart cards to reduce credit card fraud and abuse.

By late 1984, Japanese banks and chip makers entered the arena. Mitsui Bank and Toshiba Credit ran experiments in smart card banking and shopping, while Seibu Bank and Sante Systems distributed 12,000 medical cards to 300 hospitals. In May 1985 the Mirai Card Project, consisting of 1,600 monitors in eighty retail locations and nine automated teller machines, was established by Matsushita, NEC, Dai-Nippon Printing, Toppan Printing, and twelve other major corporations to test cashless smart cards. This ambitious experiment included five city banks, two regional banks, one long-term credit bank, a value-added network (VAN) to distribute transaction data between retailers and banks, and a satellite trial in the Tokyo residential area of Shirogane. By the end of 1985, there were forty major tests and over fifty smaller tests being conducted by eighty-one companies nationwide—the most intense smart card effort in the world. The Japanese zealousness was understandable. The influential Electronic Industries Association of Japan (EIAJ) forecasted that the market for smart cards and related equipment would reach a staggering $33 billion by 1995.

The smart card is rapidly becoming a Rorschach test for Japanese creativity. With their usual gusto, the Japanese are coming up with a wide variety of smart cards; just a few examples are entry keys to telephone networks and corporate data banks, biofeedback health cards, and golf cards. Memory cards are being used for data storage for desktop and laptop computers. In 1987 Toshiba and Yukiguni Yamato General Hospital began an experimental medical information system, which is regarded as a model for a smart card–based national health insurance system. On the lighter side, Japanese software makers have created singing birthday cards and smart cards capable of running computer games. The lotus blossom (MY) method described in Chapter 8 is clearly very influential.

From early trials, it appears there are no limits to the types of smart cards that can be developed. One incident, however, has stuck firmly in the minds of Japanese smart card makers. In 1985 NTT Corporation issued paper telephone cards printed with pictures of famous rock singers to encourage young people to use the cards. Initially, a limited number of cards were issued to test the market. To the surprise of NTT and other companies, these scarce phone cards became hot collector's items, selling for up to $1,200. NTT has also discovered that its paper telephone cards printed with beautiful scenes and action sports shots are being collected like stamps. It will not be long before Japanese companies sell smart cards as electronic stamps and baseball cards.

Creating Electronic Dictionaries

Behind Sharp's Wizard pocket computer and the smart card boom is a concerted effort in Japan to develop electronic dictionaries of scientific and technical terms, foreign languages, and international business. In 1986 eight Japanese computer makers formed a joint venture, the Japan Electronic Dictionary Research Center, to develop a 900,000-word dictionary for fifth-generation computers. Four dictionaries are being developed: one of 300,000 basic words, one with 600,000 technical terms, a conceptual structure dictionary, and a conceptual description dictionary. The eight-year project is budgeted at $166 million, but may be expanded to $275 million. Participating members are Fujitsu, Hitachi, Matsushita Electric, Mitsubishi Electric, NEC, Oki Electric, Sharp, and Toshiba. Their goal is to develop an electronic dictionary data management system, using up to one hundred 32-bit workstations linked by local area networks (LANs). The consortium is working closely with NTT Corporation and MITI's Fifth Generation Computer Project. Electronic dictionaries are being put into memory cards for pocket computers.

The TRON Project—A New Approach to Computing

Perhaps one of the more original computer projects being pursued in Japan today is the TRON (The Real Operating Nucleus) Project, which was conceived in 1984 by Professor Ken Sakamura of Tokyo University. Based on input from Japanese computer makers and users, the TRON Project is aimed at developing a truly Japanese open-architecture computer operating system, complete with software support, new computer designs, and a family of advanced chips. The underlying goal of the project is to help Japanese companies reduce their dependency on U.S. chip makers by developing proprietary 32-, 48-, and 64-bit microprocessors and operating systems for high-end workstations and personal computers. The TRON

Project is Japan's bid to become a major player in the battle for next-generation computers in the 1990s.

The stimulus for the project came when U.S. companies such as Intel, AMD, Motorola, and National, in an effort to avoid being deluged by look-alike products, refused to license their 32-bit microprocessors to Japanese companies. The resulting U.S. dominance of this critical technology led to chronic shortages of 32-bit microprocessors in Japan; these shortages have handicapped Japanese computer makers, especially during periods of rapid growth. Now, with the TRON Project, the Japanese hope to crack this remaining bastion of U.S. semiconductor strength. But unlike U.S. chip makers, participants in the TRON Project are not developing 32-bit microprocessors in a piecemeal fashion. Instead, they are working closely with computer users to determine their overall system needs and applications so as to develop an integrated set of computer systems, software, and chips to support these applications. The TRON Project is thus driven by a consortium of chip makers and chip users, making it a formidable example of Japanese collaborative research in action.

In June 1986 eight Japanese companies (Fujitsu, Hitachi, Matsushita, Mitsubishi, NEC, NTT Corporation, Oki Electric, and Toshiba) formed the TRON Association, which is headed by Hitachi's Kazuo Kanahara. At the beginning of 1989, the association had attracted over 125 members, including foreign companies such as IBM and Motorola. As with all Japanese joint projects, research topics were assigned to the original eight participating members, which are developing the topics on four computer operating systems, plus the TRON chips to support these systems. These four systems will be compatible with each other and will be linked by a common office or factory network (see Figure 11–1). MTRON (Macro TRON) will connect intelligent machines that feature sensors, memory chips, and embedded controllers, such as copiers and laser printers, facsimile machines, videophones, kitchen appliances, air conditioners, televisions, VCRs, and stereo receivers—all areas of Japanese industrial strength. BTRON (Business TRON) will be a common operating system for future business computers and worksta-

FIGURE 11–1
The TRON Project

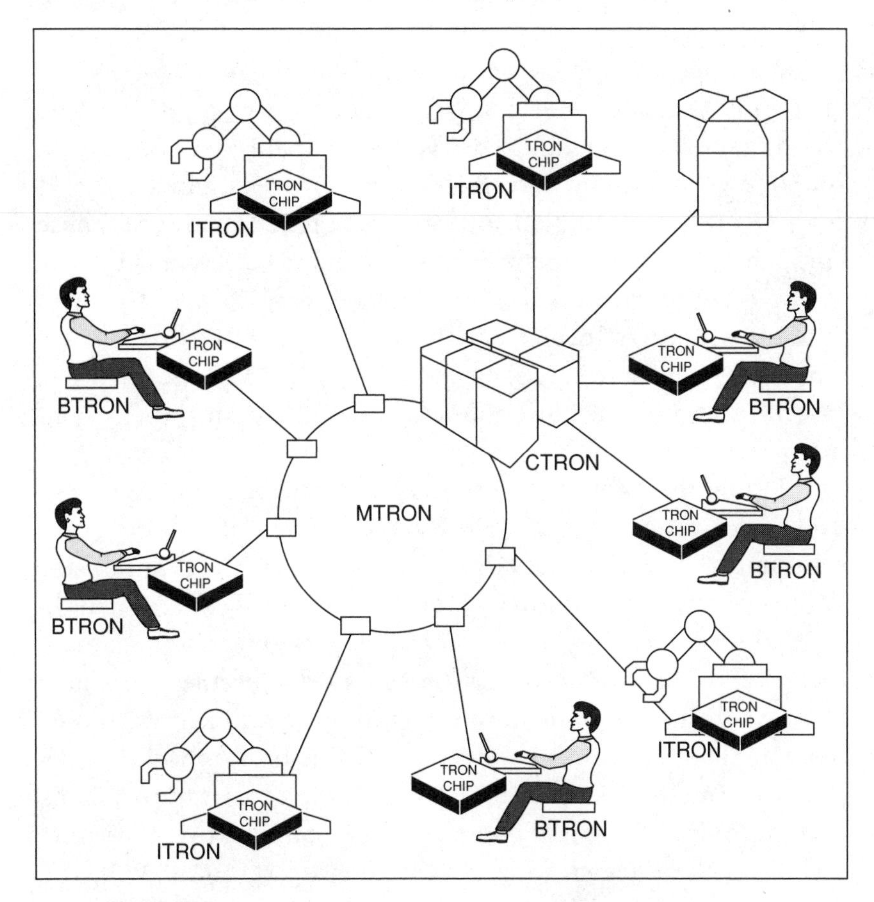

SOURCE: The TRON Project.

tions, as well as for computer-aided instruction in the schools. ITRON (Industrial TRON) will be used in embedded industrial systems, such as intelligent robots and automated inventory controls. CTRON (central TRON) will act as a communications operating system for computer networks. At its core, each TRON system features a powerful TRON chip, which is surrounded by three "layers" of operating systems for instruction processing, human-machine interface, and commercial applications. Thus, TRON is a complete operating package.

In essence, the strategy of the TRON Project is based on

the game of Go, in which one does not attack one's rival head-on but conquers by surrounding him. TRON members hope to achieve this computer victory in several ways. First, by developing an open-architecture operating system—in effect a sumo ring—TRON will force U.S. chip makers to compete directly with Japanese makers, thereby reducing Japan's dependency on one or two dominant U.S. 32-bit microprocessor makers. Second, TRON's multilingual software is designed to give Japanese companies an advantage over monolingual U.S. systems by providing software in French, Spanish, Chinese, Korean, German, Arabic, and other foreign languages. Third, TRON's focus on chip users builds on Japanese strengths in copiers, printers, factory automation, automobiles, audiovideo equipment, and many other fields.

TRON members are designing computers for a variety of users, including the deaf, blind, handicapped, bedridden, and aged, more and more of whom are looking to computers for their economic livelihood. Moreover, the physically handicapped are increasingly viewed by Japanese electronic companies as a potential computer market because of their employability in the software programming and service industries. In 1988 MITI and the Electronic Industry Promotion Association formed a committee to research the goals and problems of developing personal computers for handicapped people, including paraplegics who cannot operate standard PCs, the hearing-impaired who cannot hear beeps or other sounds, and epileptics who have problems viewing computer screen cursors.

Matsushita Communications Industrial Company is not waiting for MITI's project. In late 1988 its Central Research Laboratory developed a speech training system that allows deaf people to correct and improve their speaking ability. The $37,000 computer uses five types of sensors to detect and analyze the sounds, pitches, and tongue configurations produced by deaf speakers. The National Rehabilitation Center is developing these computers for use by the 700,000 speech-impaired persons in Japan.

These specialized computers will have broad market applications. Professor Sakamura describes how some of them will be based on current technology:

> The technology for producing voice reading of text for those with vision impairments, for example, is the same as that now used in newspapers and publishing companies to proofread text. Keyboards designed for handicapped people who cannot fully control their hand movements apply equally to the progressive aging of computer users, which must inevitably be taken into account in the future. Moreover, such keyboards would no doubt have the further advantage of being usable even in an environment subject to strong vibrations.

Professor Sakamura has proposed two design innovations for future laptop computers. One is an ergonomically designed keyboard that fits the contours of the hand and the natural extension of the fingers at rest (see Figure 11–2). By designing the keyboard to fit the hand, the TRON Project hopes to increase typing speeds and overcome the natural cramping that occurs using existing keyboards. Also, Japanese users will be able to write *kanji* characters and notes directly onto an electronic memo pad for entry into the computer system. These electronic memo pads will eventually become stand-alone "computer notepads" for entering data out in the field, such as for professional sports training, agricultural monitoring, and traffic accident reporting.

Despite skepticism in many quarters, the TRON Project is making rapid progress. Since 1986, the TRON Association has sponsored numerous programs and exhibitions to popularize the concept and attract new members. In 1988 the TRON Electronic Prosthetics Symposium was held to discuss problems the handicapped face in using computers. Another symposium, "Japanese Culture and TRON," tapped the rich traditions and sharp insights for which Osaka residents are famous. This was followed by an international TRON symposium in Tokyo.

Fujitsu, Hitachi, and Mitsubishi laid the commercial

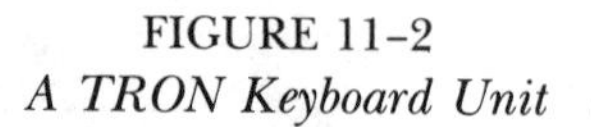

FIGURE 11-2
A TRON Keyboard Unit

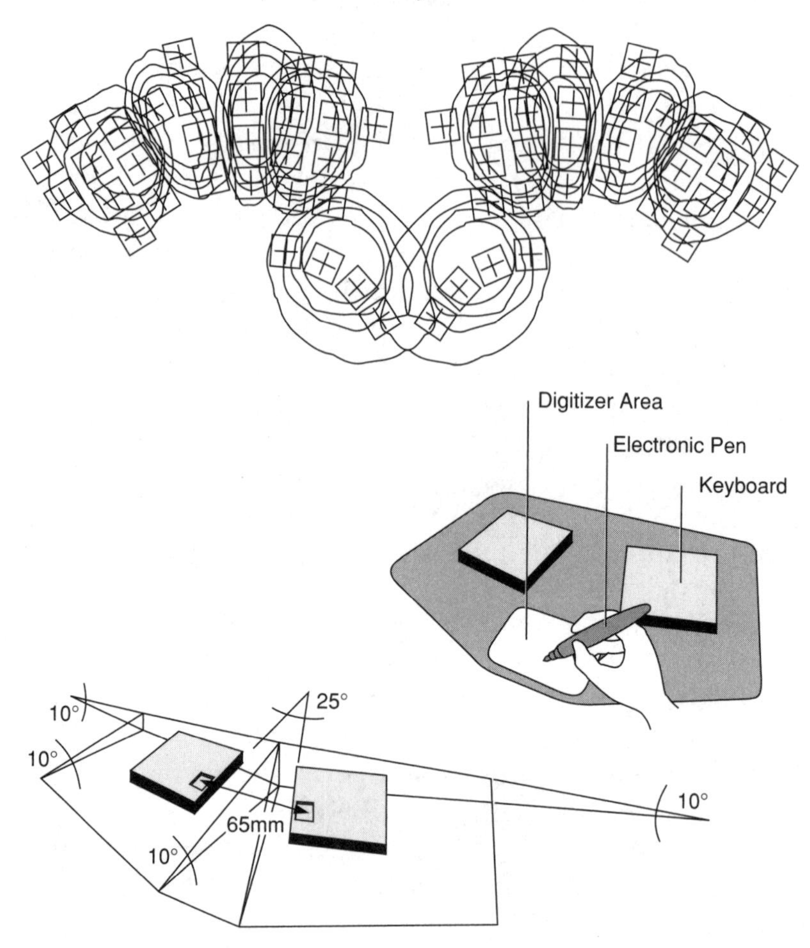

SOURCE: The TRON Project.

groundwork by jointly developing the G-Micro series of 32-bit microprocessors for personal computers, workstations, and minicomputers. They are also developing new TRON architectures, UNIX operating systems, and C-language software. MITI's Computer Education Center has chosen BTRON software for personal computers to be used in 40,000 Japanese elementary and junior high schools by 1992. (NEC, which has the dominant share with 70,000 to 80,000 personal computers already in public schools, persuaded MITI to also adopt its MS-DOS system for educational computers.) In addition, NEC, Hitachi, Fujitsu, and Mitsubishi Electric have developed commercial ITRON systems for factory automation.

Thus, the TRON Project is forging ahead and promises to develop some novel computer innovations in the 1990s.

The Supercomputer Race—Looking for Pure Speed

During the early 1980s, Japan's Fifth Generation Computer Project grabbed headlines around the world because of its ambitious goal of developing next-generation computers. But the immediate challenge to U.S. computer superiority lies in the heated race for the ultimate supercomputer—the ultrafast "number cruncher" that will enable scientists and researchers to conduct billions of computations per second. Unlike fifth-generation computers, which are still in an experimental phase, supercomputers are a reality. Since Cray Research of Minnesota introduced the Cray I in 1976, the United States has been the leader in powerful supercomputers, which form the backbone for our advanced scientific research and military defense computer systems. But thanks to support from MITI's Supercomputer Project, Fujitsu, Hitachi, and NEC have quickly caught up. During the 1980s, Japanese supercomputers became faster and extremely powerful; in the 1990s, they will challenge the technical superiority of U.S. companies.

Supercomputers are the "Formula One" race cars of the computing world. Still too expensive to be used widely in business, they are nevertheless critical for maintaining competitiveness in science and technology in which raw computing speed and applications software determine the scope of research. Using sequential (von Neumann) or parallel processing architecture and ultrafast circuits, supercomputers are capable of solving complex problems where simulation models require massive computing power, as in atmospheric studies, fusion energy research, and aerodynamic design. Existing supercomputers can tackle these problems, but they still take too much time for many applications. In order to accelerate the pace of basic research in Japan, MITI has targeted supercomputing as a strategic industry deserving government support.

In July 1980 MITI announced the National Scientific Computing Project—better known as the Supercomputer Project—whose goal was to develop by 1989 an ultrafast machine capable of ten Gigaflops (a Gigaflop equals one billion floating point operations per second: one floating point operation is a flick of a single circuit in the computer's logic network). The fastest existing supercomputers today have a capability of three to four Gigaflops. In December 1981 six of Japan's top computer makers—Fujitsu, Hitachi, Mitsubishi, NEC, Oki Electric, and Toshiba—formed the Scientific Computer Research Association to work with MITI on the project, which got underway in January 1982. Headquartered in downtown Tokyo, the project conducts joint research at MITI's Electrotechnical Laboratory in the Tsukuba Science City, northeast of Tokyo, but most of the research occurs in the corporate laboratories. The nine-year project is budgeted at $150 million, of which half comes from MITI and half from the six companies.

Due to Japan's lag in parallel processing and complex software programming in the early 1980s, the Supercomputer Project spent the first three years (1982–85) on new device technologies to create high-speed memory and logic. The goal was to design central processors capable of 40 to 100 megaflops (one million floating point operations per second).

To achieve this goal, the six member companies were assigned to four high-speed device teams (see Figure 11–3). NEC and Toshiba headed work on gallium arsenide (GaAs), a silvery, bluish-white compound metal that conducts electrical impulses five to six times faster than silicon at room temperature. All six participating companies have developed GaAs devices and have mass-produced commercial versions.

A dramatic new technology is the high-electron mobility transistor (HEMT), an aluminum-doped GaAs superlattice device proposed in 1969 by Leo Esaki and Raphael Tsu of IBM's Yorktown Heights research laboratory. Fujitsu has championed HEMT technology and has introduced several working proto-

FIGURE 11–3
The Supercomputer Project, 1981–89

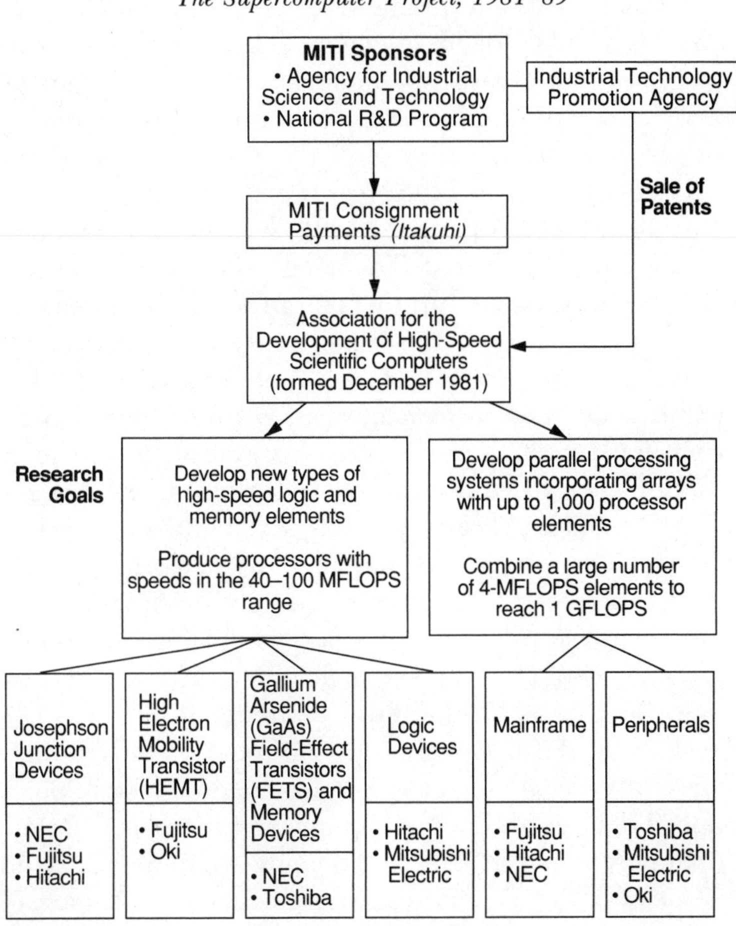

SOURCE: Dataquest, Inc.

type memories, but HEMT is likely to take a backseat to less complex GaAs devices.

Another back-up technology is Josephson junction devices (first discovered by Brian Josephson of Bell Labs in 1962), which consist of superthin films sandwiched between strips of superconducting materials (usually niobium-based compounds) that completely lose their electrical resistance when cooled at cryogenic temperatures near absolute zero (−273 degrees Celsius). These devices are still in an experimental phase and will play a limited role in next-generation supercomputers until improvements are made in fabrication, packaging, testing, and liquid helium cooling techniques.

In 1987 MITI added high-temperature superconducting ceramics as a fourth research theme after IBM Zurich made its Nobel Prize–winning discovery. But ceramics-based superconductors are still rudimentary devices and may not be commercially feasible for ten to fifteen years.

During the second phase (1985–89), the Supercomputer Project developed distributed parallel processing systems that incorporated arrays with up to 1,000 processor elements. Currently, most Japanese supercomputers use scalar processors that calculate only one instruction at a time, requiring extremely fast circuits for high performance. By contrast, vector processors can simultaneously perform calculations on data formatted in arrays. Generally, the more data vectorized, the faster the operation. Parallel processing involves using two or more processors (scalar or vector) to simultaneously perform operations in a single task. Supercomputers must use high-speed vector processors with slower scalar processors to divide tasks for optimal use in a parallel processing machine.

Since its inception, the Supercomputer Project has cooperated with the Fifth Generation Computer Project, which is also developing parallel processing systems. Since 1985, the high-speed devices and new computer architectures developed by the Supercomputer Project are being combined to build a large-scale parallel system incorporating many 40-megaflop processor arrays to reach the project's goal of ten Gigaflops.

What features will MITI's supercomputer offer? According to MITI's Electronics Policy Bureau, the new supercomputers will be user-friendly so they can be run by scientists unfamiliar with programming or machine languages. For general scientific and industrial research, the computers will be adaptable to various data bases and reporting formats. In addition to today's large mainframes, project researchers are trying to design a miniaturized supercomputer—or desktop supercomputer—that will be reliable and inexpensive to produce.

How successful has the Supercomputer Project been? So far, Japanese supercomputer makers have achieved parity in raw machine speed, but still lack software for vector processing. In 1983 Hitachi introduced its S-810/20, which runs at 630 megaflops, comparable to the Cray X-MP48. That same year, NEC introduced the SX-2, which runs at 1.3 Gigaflops, the first Japanese supercomputer to exceed one billion floating point operations per second. In late 1988 Fujitsu raised the ante by announcing its VP-2600, which was topped in April 1989 by NEC, who stunned the industry by announcing its SX-X series supercomputer that will run at the blistering speed of twenty-two Gigaflops—outpacing the Cray-3, which will run at sixteen Gigaflops (see Table 11–1).

According to Sidney Fernbach, a former scientist at the Lawrence Livermore National Laboratory who has evaluated Japanese supercomputers, "The most interesting fact about these [Japanese] systems was that the software was highly

TABLE 11–1
Commercial High-Performance Scientific Supercomputers (over 500 megaflops)

Company	Model Name	Date	Maximum Speed (mflops)	Central Memory (megabytes)
Cray	X-MP48	1985	840	n/a
	Cray-2	1985	1000	256
	Cray-3	1990	16000	n/a
Denelcor	HEP-2	1985	1000	2000
Fujitsu	VP-200	1982	530	256
	VP-400	1985	1070	256
	VP-400E	1987	1700	n/a
	VP-2100	1988	500	32–1000
	VP-2200	1988	1000	64–1000
	VP-2400	1988	2000	128–1000
	VP-2600	1988	4000	128–2000
Hitachi	S-810/20	1983	630	256
	S-810/60	1987	1500	n/a
	S-820/80	1987	2000	n/a
NEC	SX-1	1983	570	128
	SX-2	1983	1300	256
	SX-3	1989	5000	n/a
	SX-X	1990	22000	64–2000

SOURCE: Dataquest, Inc.

advanced, well optimized, and automated." U.S. supercomputer companies currently have a lead over Japanese companies in writing complex software for vector processing, but that lead is narrowing. The Japanese focus on ultrafast devices—their area of strength—is already paying off. As Dr. Norman H. Kreisman, adviser for international technology at the U.S. Department of Energy, notes:

> U.S. companies are using components which are sometimes inferior to those available to the Japanese; [U.S.] speed advantages come, for now, from advanced machine architecture, such as parallel processing. The Japanese are moving into parallel systems. When this happens, the game may revert to one of component speeds and competitive software.

John Richardson, chairman of the Institute of Electrical and Electronics Engineers' (IEEE) supercomputer advisory board, is more pointed: "The Japanese systems are already a match for U.S. systems and soon could surpass them in performance."

Because of Japan's rapid gains in supercomputer technology and the Japanese government's "buy Japan" procurement policies, market access for U.S. supercomputer makers has become a major issue in Japan–U.S. trade talks. In 1988, after intense negotiations, the Japanese government agreed to implement new supercomputer procurement rules to make its bidding process more "transparent." The agreement provides that (1) all procurement plans by Japanese government agencies must be announced publicly; (2) companies involved in developing procurement specifications are to be banned from bidding; (3) foreign suppliers must be given ample time to submit bids, specifications, and other materials; (4) Japanese procuring agencies must consider technical factors as well as cost in evaluating bids; (5) the procuring agency will publicly announce each process leading to the winning bid and will explain to losing companies why their bids were not accepted; and (6) a complaints section will be established by each procuring agency.

While these rules provide greater transparency, the question of Japan–U.S. reciprocity is still a contentious issue because of the mixed economic and military objectives of the United States. In the early 1980s, for example, the Pentagon opposed the sale of Cray Research's advanced supercomputer to NTT and other Japanese agencies because of concern over the loss of valuable technology to the Eastern bloc. Later, the U.S. Department of Commerce criticized the Japanese government for not buying U.S. supercomputers. Similarly, the Pentagon recently discouraged MIT from buying a Japanese supercomputer, despite U.S. government demands for purchases of U.S. supercomputers by Japanese government ministries to maintain reciprocity in our trade relations with Japan. No wonder Japanese and Americans alike are confused by the conflicting signals coming out of Washington.

As long as Japan trailed the United States in supercomputer technology, this "buy America" policy might have made sense. But as Japan achieves technical parity and, in areas such as commercial high-speed chips, technical superiority, such a policy will ultimately prove self-defeating and counterproductive. Maintaining global competitiveness requires access to the best technology available, whatever its origin. Handcuffing the United States with technological protectionism will only undermine U.S. industries and universities. If Japanese researchers have access to the fastest supercomputers, the United States will lose by default. What the United States needs is not more protectionism but a coherent supercomputer policy and a development program designed to overcome Japan's current advantage.

Has the Fifth Generation Computer Project Failed?

Perhaps the most controversial Japanese effort has been MITI's ambitious Fifth Generation Computer Project, which was organized in 1982 to develop state-of-the-art technologies for next-generation computing. Originally budgeted at $450 million, the project represents the first radical departure from existing computer technology in over thirty years. Unlike current von Neumann computers (named after computer inventor John von Neumann), which can only process alphanumeric data sequentially, the fifth-generation computer will use vast arrays of parallel processors to handle symbolic logic, images, graphics, speech, and other types of data that require large amounts of computing power. In the past, von Neumann computers were adequate for most computing operations, but the information explosion has rendered them inadequate, especially for the Japanese, who are more oriented to image data.

In Japan there is enormous interest in developing computers better able to handle Japanese styles of thinking and expression. Their *kanji* writing system is ideographic, not phonetic like Western alphabets, and their thinking patterns are not historically based on linear, Descartian rules of logic. As J. Marshall Unger, author of *The Fifth Generation Fallacy,* observes:

> The Japanese commitment to strong artificial intelligence research is intimately related to the nature of the Japanese writing system. Unless a new, fundamentally different kind of computer can be built, the inefficiency of using traditional script in computer environments will become intolerable as the scope and number of computer applications grow.

Indeed, the Fifth Generation Computer Project appears to be Japan's attempt to overcome the tyranny of alphanumeric data entry and Western logic rules, which have dictated computer architecture and software programming for decades.

To achieve these goals, the Fifth Generation Computer Project has chosen to develop a new computer architecture combining a large knowledge base, a problem-solving and inference system, and user-friendly terminals. The knowledge base would consist of large libraries of images, graphics, voice messages, and other symbolic data inputted through the use of speakers, optical readers, sensors, and other devices. The data would be represented in the form of "objects," or clusters of attributes, which would be related to other bits of data by "links," or symbolic references. The knowledge base would consist of a hierarchy of objects and links tied together by rules consisting of "if . . . then . . . " statements, not necessarily following Western rules of logic. For example, a person might ask if a catfish can be used to make *sashimi;* the computer would reason: "Only saltwater fish can be used for *sashimi,* and catfish is not a saltwater fish; therefore, catfish should not be used for *sashimi.*" A sophisticated knowledge base developed by Japanese *sushi* makers might include highly subjective characteristics, such as texture, coloration, sheen, smell, or softness. This knowledge would be stored in a large file, called a "relational data base," that could be automatically updated by knowledge-base management software.

The second component of the Fifth Generation Computer Project is the problem-solving or inference subsystem. The Institute of Next-Generation Computer Technology (ICOT), which is administering the program, has chosen PROLOG (Programming in Logic) for the machine language over LISP (List Processing), which is popular in the United States, because of PROLOG's pattern-matching and nondeterministic logic structure. Although Americans question the practicality of PROLOG because data is often presented in an arcane format and its automatic theorem-proving process prevents programmers from understanding the logic steps, ICOT is pursuing it because Japan has little invested in either language and PROLOG offers greater flexibility in representing the knowledge base. ICOT has two goals for the inference system. The first goal is to develop a one-user PROLOG workstation capable of performing one million logical inferences per second (LIPS), where one inference requires 100 to 1,000 instructions. The final goal is to build a computer capable of executing one billion LIPS.

The third component is user-friendly computer terminals that make it easy to input voice, graphic, image, or handwritten data. Existing terminals are basically modified typewriter keyboards that require users to be proficient in typing. The fifth-generation computer will use speakers, question-and-answer systems, optical scanners, touch screens, "mice," video cameras, cameras, phonetic typewriters, and other nontyping means for inputting data, making computers less like office machines and more like household appliances.

ICOT initially focused on three basic applications for the fifth-generation computer. Japanese–English translation software is a major goal because of the linguistic barriers between Japan and the rest of the world. ICOT's original goal was to develop a prototype multilingual translation system with a 100,000-word vocabulary and the ability to translate, edit, and print with 90 percent accuracy. A second application would be a small interactive multimedia library, with information-gathering capacities 100 times greater than existing systems, for specialized fields such as medical diagnosis, computer-aided design (CAD), and equipment repair. Finally, ICOT planned to introduce nearly automatic software creation with little need for computer literacy on the part of the user. Automatic software generation was originally part of the Fourth Generation Computer Project (1975–80), but MITI encountered difficulties with it and pushed it into this project.

ICOT envisioned that the fifth-generation computer would be developed over ten years in three phases. In phase I (1982–84), which was budgeted at $70 million, ICOT researchers were trained on PROLOG programming, and a single-user sequential PROLOG workstation and software for expert system applications were developed. In phase II (1985–88), budgeted at $213 million, parallel processing and knowledge base subsystems were merged, with work continuing on a new inference subsystem. In phase III (1989–92), all three subsystems will be merged into a single fifth-generation computer. Each company was assigned a specific task:

Mitsubishi/ Oki Electric	Personal sequential inference computer model
NEC	Sequential inference machine
Toshiba	General-purpose parallel computer for knowledge bases and relational data bases; local area networks (LANs)
Hitachi	Hierarchical memory subsystems

NEC/Hitachi/ Fujitsu	Parallel inference and knowledge base machines
Tokyo University	Parallel inference machine for PROLOG machine
Mitsubishi	Bit-mapped display terminals and "mouse"

Has ICOT made much progress? Since the mid-1980s, there has been much speculation as to whether ICOT has achieved its goals. Some observers have claimed that the Fifth Generation Computer Project is a colossal failure; others dismiss it as not very innovative. While it is true that ICOT has not achieved all of its goals, many of these criticisms are based on two basic misconceptions about the project.

"ICOT Does Basic Research"

Probably the biggest misconception about the Fifth Generation Computer Project is that ICOT is primarily devoted to pursuing basic research in AI, relational data bases, and parallel processing. Edward A. Feigenbaum and Pamela McCorduck contributed to this view in their book *The Fifth Generation:*

> Their [the Japanese] goal is to develop computers for the 1990s and beyond—intelligent computers that will be able to converse with humans in natural language and understand speech and pictures. . . . The Fifth Generation will be more than a technological breakthrough. The Japanese expect these machines to change their lives—and everyone else's.

Although ICOT conducted a worldwide literature search, it did not pursue basic research during phase I. In late 1987 the Japanese Technology Evaluation Program (JTECH), a team of top U.S. computer experts, visited ICOT and major Japanese computer companies to assess the status and long-term direction of Japanese research in advanced computing. In the panel report to the National Science Foundation, Marvin Denicoff of

Thinking Machines Corporation observed: "The USA JTECH panelists unanimously concluded that the Japanese effort to date has produced no fundamental advances."

Joseph Goguen of SRI International and Carl Hewitt of MIT, members of the JTECH team, explain this lack of fundamental research:

> ICOT is not doing basic research, and is not supposed to be doing basic research, either in artificial intelligence or in any other field. Rather, they are doing innovative long-term applied research. This means that they want to build prototypes that demonstrate specific progress toward difficult long-term goals. *ICOT is focused on developing innovative technology, rather than basic research* [emphasis added].

"Where Are the New Products?"

Some believe that since ICOT has not produced any commercially viable products, the project is a failure. This misconception contradicts the first one: how could ICOT be focused on basic research and product development at the same time? Moreover, it naively assumes that commercially viable products can spring from basic research within a short period of time.

It is still too early to tell whether commercially viable products will emerge since the Fifth Generation Computer Project is just entering its commercialization phase. Generally, the locus of MITI projects shifts to the private sector during the last phase. Research activities "disappear" from sight into the corporate laboratories, leading many foreign observers to conclude that a project has failed. In fact, companies jealously maintain secrecy and guard their proprietary technology since they are closer to product introduction. Thus, when the JTECH team visited ICOT during phase II, Goguen and Hewitt prematurely concluded:

> The Fifth Generation Project is not trying to produce prototypes for commercial products, but rather

> research prototypes . . . ICOT is not doing short-term development, with specific low-risk goals, they are doing innovative long-term applied research, and they know that this is high risk, and that a lot of blind alleys will have to be explored.

In fact, the project had not yet entered its commercialization phase and commercial prototypes were unavailable.

If the Fifth Generation Computer Project cannot be evaluated from the standards of either basic research or commercially viable products, how can ICOT's success or failure be measured? There are several ways of evaluating ICOT—by looking at its social and business impact, and by assessing its technical achievements.

Goal #1: To reverse Japan's image as a copier. Perhaps the most important goal of ICOT's director, Kazuhiro Fuchi, has been to erase Japan's image as a copycat. ICOT has succeeded admirably in commanding interest and respect as an equal partner in the international research community. Westerners no longer deride Japanese computer research, but carefully watch each new technical development and product innovation.

Goal #2: To keep Japanese researchers abreast of worldwide research trends in AI, parallel processing, and other leading-edge topics. In this respect, ICOT has succeeded brilliantly by acting as a funnel for information on worldwide research trends. When the project was founded in 1982, Japanese computer researchers were not welcome in foreign laboratories around the world because of the IBM "sting" operation: Hitachi and Mitsubishi managers were caught illegally buying proprietary IBM information. To overcome this technology boycott, ICOT invited respected experts to present technical papers in Tokyo and attracted over 1,000 leading AI researchers. In this way, ICOT was able to gather key information and identify its worldwide competitors without venturing abroad.

Goal #3: To develop creative new AI and parallel processing technologies. ICOT has been instrumental in forging the new AI technologies now being pursued by corporate laboratories and other government projects. Marvin Denicoff of Thinking Machines Corporation, a member of the U.S. JTECH team, observes:

> The Fifth Generation Project has been heavily involved with speech and image processing and has produced tangible results in such fields as language translation, speech and character recognition, document and mail processing, and expert systems. . . . There is no disagreement in recognizing that these Fifth Generation results are state-of-the-art and compare favorably with U.S. products, even teaching us a lesson in the speed of development and smooth industrial coupling of these commercially-directed efforts.

What are some of the project's technical achievements? ICOT developed sequential, single-processor machines during phase I. In 1985 Mitsubishi Electric introduced a personal sequential inference (PSI) machine, Melcom PSI-I, the first Japanese computer to commercialize PSI technology, which runs on PROLOG and has a speed of 40,000 LIPS. It was followed in 1987 by the PSI-II, which is three times faster and half as expensive. The Mitsubishi machines are being marketed for use in automatic diagnostic systems and decisionmaking systems in chemical, steel, food processing, and transportation firms.

One technical breakthrough was made in 1985 by Kazunori Ueda (assigned to ICOT by NEC); he developed the Guarded Horn Clause (GHC) base language, a follow-on to Concurrent PROLOG. The advantages of this language is its simplicity and ability to support higher level programming languages.

At its 1988 symposium in Tokyo, ICOT displayed Multi-PSI 2, which is capable of understanding sixth grade–level Japanese-language textbooks, and Multi-PSI 3, a system that can understand and respond to questions. In 1989 ICOT began developing a parallel inference machine operating system (PIMOS) and was planning to unveil a 128-processor logical inference machine (LIM) capable of processing more than 20 million LIPS. The machine will feature clusters of eight processing elements each. A LIM would connect sixteen clusters on a twisted-pair LAN to achieve the 128-processor level.

Goal #4: To encourage Japanese corporations to pursue more creative research in AI and other underexplored fields. Since 1982, there has been a research boom in AI, parallel processing, automated language translation, and other topics targeted by ICOT. Hitachi, Melcom, NTT, Toshiba, and Kyoto and Tokyo universities are building very fast symbolic programs that handle Japanese characters. New AI software languages are sprouting up, such as Kyoto Common LISP, KL-2, TAO, SUPER-BRAINS, ES/Kernel, PROLOG-KABA, and UTILISP. And computer makers have developed symbolic processing prototype machines, such as PSI-II (Melco/Oki/ICOT), IP 704 LISP board (Toshiba), the ELIS LISP machine (NTT), and the NEWS workstation (Sony). There is so much activity that the *Japan Economic Journal* began publishing *Nikkei AI* magazine in mid-1985 to focus the intense interest in this emerging field. Japanese companies are already using expert systems, but most AI software will be used in-house. ICOT and the government-sponsored AI Center forecast a Japanese AI software market of $32 billion in 1995, of which $4.4 billion, or 13.8 percent, will be merchant sales.

Electrical power systems, factory automation, health care, and electronic banking are the first areas in which Japanese are applying AI systems. Japan's nine regional electrical power companies are introducing expert systems for power plant operations and nuclear reactor safety, while manufacturing companies are using them for electrical power monitoring. In 1987, for example, Mitsubishi Electric announced an AI system for its electricity supply routing support system. This system shows power amounts, destinations, and feeder routes used by utility companies, eliminating load-dispatching engineers through the use of inference-based data bases. Hitachi and Toshiba are developing AI systems for utility companies.

Hitachi has developed a new AI computer language, SONLI, that categorizes and manages large knowledge bases by dividing them into smaller increments by function and linking them to perform interference functions efficiently. The program runs seven and a half to twelve times faster than conventional knowledge base systems. In 1986 Hitachi and West Nippon Bank developed a high-speed compact expert system called "Eureka" to assist commercial loan operations. In 1988 Fujitsu began exporting AI products, including an expert system development tool (ESHELL), an English–Japanese/Japanese–English translation system (ATLAS), and a back-end LISP machine (FACOM). Thus, the Fifth Generation Computer Project has triggered quite a bit of research on commercial AI and expert systems, research that will increase during the 1990s when the project is over.

As the Fifth Generation Computer Project enters its final phase, ICOT aims at developing a complete parallel processing machine using 1,000 processors, complete with basic software and applications software. The biggest challenge will be to develop the software environment that can perform logical operations on a massively parallel system. To date, the natural language processing and intelligent software development have been difficult problems. But Dr. Fuchi is optimistic. "Next year will see the beginning of the final stage, the jump stage. This will be the period for in-depth research on parallel inference software. . . . We expect parallel inference to be the core of future information processing, and we believe that it will lead to a new type of computer."

The Sigma Project

The massive software development challenge of the Fifth Generation Computer Project has highlighted the need for more effective ways of writing software. According to MITI, Japanese industry may be confronted with a shortage of one million programmers by the year 2000, at the present rate, unless more programmers are trained and new software programming techniques are developed. Since the Fourth Generation Computer Project, one of MITI's dreams has been automated software development. But MITI researchers encountered numerous problems. They first pushed the research into the Fifth Generation Computer Project; now it is part of the Sigma (Software Industrialized Generator and Maintenance Aids) Project.

In April 1985 MITI launched the Sigma Project, a $210 million project to create systems for enhancing the productivity of UNIX-based software development. The goal is to quadruple productivity by developing 80 percent of software, not through manual coding, but through a national software data base. There are several related subgoals: enhancing software quality and productivity, accumulating software know-how, reducing duplicated software investments, and improving engineer training. By 1990 the Sigma Project will complete a nationwide network system that links software makers to the national Sigma Software Center. Currently, 250 companies are participating in the project, which is supervised by the Information Processing Association (IPA). In 1987 IBM Japan finally joined, after the project ordered high-performance personal computers, workstations, and Japanese-language software from Japan's "Big Six" computer makers: Fujitsu, Hitachi, Mitsubishi, NEC, Oki Electric, and Toshiba.

The Sigma Project is divided into two phases. During phase I (1985–86), fifty-nine companies developed the "Vo Document," which officially defined the content of the Sigma operating system. It is a UNIX-based system that will be supported by Japanese-language code as defined by the Japanese UNIX Council; it will have a basic section, multi-windowing, and graphics. In addition, the project developed Sigma workstation tools, monitors, LANs, hard disks, and machine interface tools.

During phase II (1986–90), the project is field-testing the Sigma operating system on large-scale computers. In early 1987 the Sigma Software Center was opened to provide users with data base and networking services. Fujitsu, Hitachi, NEC, and NTT were selected to develop five subsystems for the control center: networks (NTT), data bases (Hitachi), demonstrations (Fujitsu), electronic data processing (NEC), and development environment (Fujitsu). The DARPA Internet protocol is being used initially to link Sigma projects with overseas networks, but it will eventually be replaced by the OSI protocol.

In mid-1987 IPA installed the multi-word system into Sigma workstations. IPA's goal is to develop the Sigma standard software system "Multi-Media Word," which will be capable of handling characters, graphics, and figures. Kyoto University developed the Window System, which has powerful Japanese-language capability for use in Sigma workstations.

Software-Less Computers

Although automating software development—whether by the Sigma Project or by computer-aided software engineering (CASE)—would help boost computer sales, the ideal would be to develop computers that do not require software. In August 1988 Casio Computer announced such a computer, the Automatic Data Processing System (ADPS); it performs tasks by reading function codes accompanying the files and input data. The computer has only five commands: append, delete, correct, total, and retrieve. Casio Computer, which spent fifteen years developing the system, commercialized it in 1988.

NTT's INS Computer

While MITI has developed its fifth-generation computer, NTT is spending over $550 million to develop an advanced computer for its $120 billion Information Network System (INS). The INS computer will combine fifth-generation computer technologies (knowledge base, learning acquisition, inference mechanisms, natural language translation, speech recognition, automatic software production, and audiovideo information processing); supercomputer technologies (parallel processing and high-speed logic circuits); and optoelectronic technologies (megabit video memory chips, video processing, and optical transmission). The INS computer will merge communications and computer technologies into an image- and video-oriented system with the potential for automated voice translation and information processing. When completed, the INS computer will free Japan from the constraints of alphanumeric data entry and will allow it to build on its strengths in video technology.

The INS computer will form the information and communication node of the INS network. The INS program is divided into three phases: phase I (1984–89), development of a prototype system; phase II (1990–93), installation of a central INS computer in the NTT headquarters; and phase III (1993–2000), multiple INS computers installed in major switching nodes throughout Japan to provide distributed communications processing.

INS research began when NTT established the Base Intelligence Research Group at its Yokosuka Electrical Communication Laboratory (ECL) in 1982 to pursue fifth-generation computer research paralleling work at ICOT. MITI and NTT frequently hold joint meetings to exchange technical information. Recently, MPT modified the INS program to make it compatible with the Integrated Services Digital Network (ISDN) international standards. Since April 1988, NTT and its competitors have begun investing heavily in developing ISDN services and equipment. In 1988 alone, NTT will invest about $13.7 billion in ISDN equipment and facilities. Some of this investment will go toward modifying the INS computer for NTT's ISDN network.

Translation Phones—Overcoming the Language Barrier

As Japanese industry internationalizes, the ultimate dream of both business managers and MPT is to develop a telephone that can automatically translate spoken words between Japanese and all major foreign languages. These translation phones would not only reduce second-guessing and miscommunications with customers and overseas offices, but would also drive innovations in computer and software technologies. To promote this technology, MPT prepared a master plan in 1985 to develop a translation telex system by 1995 and a translation phone by the year 2000 at a cost of $833 million. In 1987 the Council for the Promotion of Automated Translation Telephone Systems Development issued a report outlining a future automated phone system. The council recommended developing a prototype system with limited language skills by the late 1990s, at a cost of $750 million, with the following development goals:

- 1991—A simulation model with a 2,000-word vocabulary and limited grammar and enunciation capabilities
- 1996—Slow automatic translation system with a 3,000-word vocabulary that could be used with international telephone operators
- 2001—Moderately fast automatic translation system with a 10,000-word vocabulary
- 2006—Ultimate system with a three million–word vocabulary for handling routine telephone conversations

The translation phone will feature a TV camera, monitor, and readout to display the time, day, and country of call destination. Videophones will be merged with translation phones to develop next-generation translation videophones with which callers can speak to each other over Videoman-like phone handsets.

In 1985 the Key Technology funded the Advanced Telecommunications Research (ATR) Interpreting Telephone Research Laboratories in the Osaka region. The goals of the ATR Interpreting Telephone Project, a ten-year program budgeted at about $100 million, are to develop a real-time telephone translation system and to conduct research in sight and hearing for better human/machine interface, as well as basic research in light communications. The project is studying three basic technologies that are required for automatic voice translation: sound recognition, sound synthesis, and machine translation. Under Kiyohiro Shikano, head of the Speech Processing Department, ATR has analyzed the sound waves of over 8,500 tape-recorded words spoken by professional announcers. When the three "a" sounds in the Japanese word *sazanami* (ripples) are shown on an oscilloscope, for example, each one shows a slightly different sound wave, notes Shikano.

NTT, Kokusai Denshin Denwa (KDD), NEC, Fujitsu, Oki Electric, IBM Japan, Nihon DEC, Toshiba, Sony, Hitachi, and thirty other companies are participating in the project.

The groundwork for future translation phones is being laid in natural language translation computers. In October 1988 seven top Japanese computer makers displayed a prototype computerized system that translates Japanese into four Asian languages—Chinese, Indonesian, Malay, and Thai—as well as cross-translates among these languages. The machine was the result of cooperation between the sixty-six-firm Center of International Cooperation for Computerization and the governments of China, Indonesia, Malaysia, and Thailand. Designed to handle technical documents, the machine can translate up to forty sentences and can handle about 95 percent of all translation, requiring humans only for the final revisions. An electronic dictionary used for the system contains 1,000 Japanese and Chinese words, but will be expanded to 75,000 words.

Japanese companies are also developing personal word processors featuring voice-input capabilities that will eventually be used in translation phones. Sharp, for example, has introduced a $3,500 personal word processor that accepts spoken words in complete sentences. The machine is speaker-dependent, requiring about ten minutes to register 10,000 voiced words into the system.

Thus, natural language translation phones and computers are a hot research topic in Japan. By the end of the ATR Interpreting Telephone Project in 1995, Japanese companies will introduce a variety of ISDN translation phones capable of translating voice and text.

Optocomputers—Computing at the Speed of Light

Despite advances in computer technology, all existing computers are based on one operating principle: transmitting information using electrical impulses. The supercomputer would accelerate the flow by introducing ultrafast circuits; the fifth-generation computer would employ symbolic logic and parallel processors. But both technologies are handicapped by several constraints inherent in electrically based systems. Not only do they consume large amounts of energy and require coolants to dissipate waste heat, they are extremely vulnerable to electromagnetic interference—a serious problem in high-level applications such as commercial aviation, automated manufacturing, space exploration, and military defense. Moreover, silicon chip technology is rapidly reaching physical limits in microcircuitry.

To overcome these obstacles, MITI is working with Japanese companies to create a computer based on light rather than electrical impulses. This work is proceeding in two phases: phase I (1981–86), developing factory optical sensors; and phase II (1986–95), improving optical devices for optical communication and optocomputers.

In the first phase, the Optical Measurement and Control Systems Project (better known as the Optoelectronics Project) was formed to develop optical devices for image processing and data transmission. These systems can be used for running industrial processes by remote control, as well as for image transmittal by large offices, shopping centers, and medical and traffic systems. The project focused on developing an optical system that could operate in the presence of explosives, inflammable gases and chemicals, electromagnetic induction, intense radiation, and high temperatures and humidity—conditions that cannot be safely handled by electrical systems. The prototype was a petroleum refinery near the Inland Sea.

The Optoelectronics Project involved over 100 companies that are organized into the Engineering Research Association of Optoelectronics Applied Systems. In 1981 a joint research laboratory was established within the Fujitsu laboratory in Atsugi. The project was budgeted at $150 million, and all expenses were covered by MITI consignment payments (*itakuhi*), which companies did not have to repay. Patents developed from the project were government property, but available to participating companies.

The Optoelectronics Project was a stunning success. MITI filed more than 300 patents, while Japanese companies gained world dominance in semiconductor lasers, optocouplers, photodiodes, high-output light-emitting diodes (LEDs), charge-coupled device (CCD) sensors, and first-generation optoelectronic integrated circuits (OEICs). These devices are now being used in laser and compact disks, 8mm video cameras, optical disk systems, optical communications, factory automation monitoring, remote control systems, TV animation, laser copiers and printers, point-of-sale terminals, and office LANs. By 1985 over 1,000 optoelectronic systems had been installed in Japan.

But MITI's ultimate goal is to create the dream computer—an optocomputer that would be faster and less energy-guzzling than current computers. Instead of using metal circuits, optocomputers would feature tiny lasers that shoot light signals directly to photodetectors. An optocomputer armed with these semiconductor lasers and OEICs would have numerous advantages over electrically based computers. Instead of processing numbers and letters, they could process images—the medium for the fifth-generation computer and future neural network computers. They could be used as the "eyes" and "brains" for robots monitoring and controlling industrial activities in harsh weather conditions or hazardous environments. Optocomputers could also be linked to fiber-optic telecommunications networks, enabling telephone companies to handle easily many more calls. Envisioned is the eventual merging of optical computers and communications into a single "C&C" system.

In early 1986 a Tokyo University research team under lecturer Kazuo Hotate developed a prototype optocomputer using an optoelectronic device that combines ten lenses to run complex matrix calculations simultaneously. By lining up lenses in a single line and placing screens at three intermediate points, the test machine was able to process incoming light and display the resulting calculation as an image at the opposite end. This two-dimensional optical array was Japan's first step toward an optocomputer.

In June 1986 thirteen Japanese companies formed the Optical Technology R&D Corporation to develop second-generation OEICs for optocomputers and optical communications. The participating companies include Fujikura, Fujitsu, Hitachi, Koga Electric, Japan Glass, Matsushita, Mitsubishi, NEC, Oki Electric, Sanyo, Sharp, Sumitomo Electric, and Toshiba. The ten-year project has a budget of $83 million, and MITI has lent $29 million worth of equipment at no cost to the companies currently receiving a $5 million supplement for OEIC research. In 1987 the venture opened its Tsukuba Research Center near MITI's Electrotechnical Laboratory in the Tsukuba Science City. The project goal is to develop two-dimensional optical chips, then three-dimensional OEICs for experimental optocomputers by 1995.

Research assignments are divided up among the participating companies; topics included are optical switching, optical signal processing, and optocomputer architectures. Their goal is to create in the early 1990s a high-speed optical switching computer for communications. By the late 1990s they hope to develop an optoelectronic computer—and in the early twenty-first century, a true optocomputer that will feature artificial intelligence. At each level of complexity, they will develop more sophisticated OEIC devices—one of Japan's competitive strengths. In 1987 Toshiba developed an ultrafast optoelectronic device capable of 10 × 10 parallel processing for optocomputers, enabling signals to travel 100 times faster than they do on current supercomputers. In 1988 Fujitsu developed the world's first optical memory device; called the bistable laser, it may be used in optical switching systems and optical computers. Thus, participating companies are making rapid progress toward optocomputing.

U.S. experts are becoming concerned about Japan's lead in optoelectronics and optocomputing. In February 1988 the U.S. International Trade Commission issued a report, "U.S. Global Competitiveness: Optical Fibers, Technology, and Equipment," warning that the large Japanese and European companies would shape the future of optical technologies. Although Bell Labs and RCA have developed basic optical technologies, former researchers at those companies mention that large U.S. companies have not been willing to pursue the integrated approach taken by the Japanese during the 1980s. The National Academy of Engineering echoed this warning in its 1988 report, "Photonics: Maintaining Competitiveness in the Information Age": "The development by the Japanese of the full potential inherent in photonics could threaten America's leadership in several areas" (including telecommunications, computing, optical storage and display, and optical sensors). The academy recommended the creation of a national photonics project focusing on automotive electronics, supercomputers, optical links, and optoelectronic integrated circuits.

Neurocomputers

In June 1987 the first annual conference on neural networks was held in San Diego, California, and attracted a crowd of 1,500 scientists and researchers, including many from top U.S. start-up companies as well as from Japan. All attendees were intent on understanding a renewed technology that had been pioneered by Warren S. McCulloch of the University of Illinois and Walter Pitts of the University of Chicago in 1943. The first neurocomputer, the Perceptron, was built by Frank Rosenblatt at Cornell University in 1949, but research was discontinued in 1969 when Marvin Minsky and Seymour Papert of MIT suggested that the Perceptron neurocomputer was unsuitable for artificial intelligence. In 1982 Cal Tech professor John J. Hopfield broke open the field again by suggesting that certain neural networks could solve the classic "traveling salesman" problem (finding the shortest path through a group of cities), which had resisted conventional computers. This breakthrough, which was as dramatic as the recent discovery of high-temperature superconductors, triggered a worldwide race to develop the next-generation neurocomputers.

What is a neural network? It is basically a network of artificial neurons that consists of various input and output circuits. Figure 11–4 shows the basic structure of a neurocomputer; it is divided into an input layer, a hidden or inner layer, and an output layer of neurons (each neuron is represented by a cir-

FIGURE 11-4
Structure of a Neurocomputer

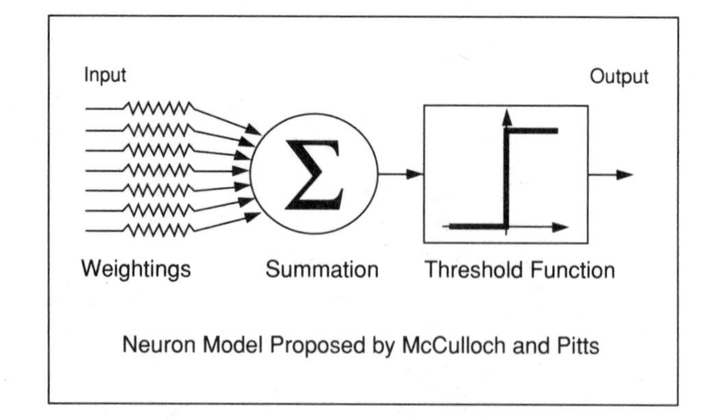

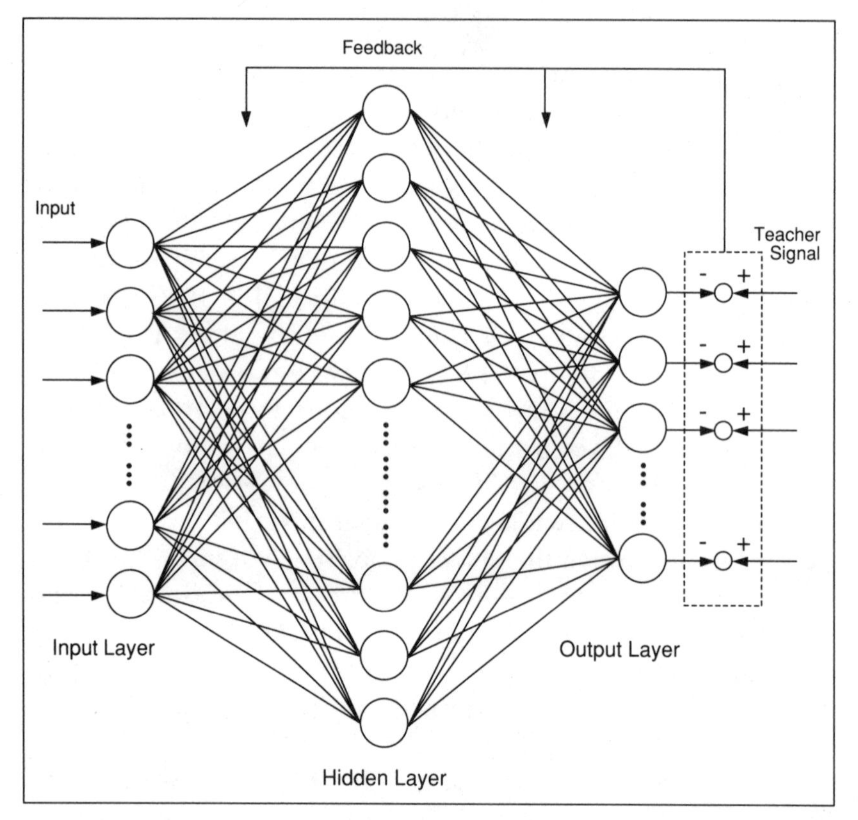

cle). Incoming signals enter the input layer, which transfers them to the inner layer, which weighs and sums them, then sends the results to the output layer. The advantage of artificial neurons is that they can handle a variety of different inputs, such as handwriting or voice data, and they can be "trained." A neural network allows the threshold levels and weightings for each layer to vary, so the network can optimize the weightings by itself. Essentially, the neurocomputer can "learn" by experience, instead of having to be preprogrammed.

Simple neurocomputers are already capable of recognizing voices, shapes, images, handwriting, and other nondiscrete types of data. Moreover, they can be developed within a few months; developing expert systems and debugging their software takes several years. Because of these enormous advantages, Japanese computer makers are excited about their prospects for sixth-generation computing.

In April 1989 MITI began a neural network computer program that will be in full swing by 1992 and will last ten years. The project will develop neural network systems and devices capable of massive parallel processing and fuzzy logic for sophisticated voice, image, and pattern recognition. Research will focus on five areas: large-scale emulators, algorithms, neural network devices, pilot neural network computer systems, and commercial applications. For 1989, MITI asked for only $183,000 start-up capital, but funding will increase over time. According to Yuji Tanahashi, the director-general of MITI's Machinery and Information Industries Bureau: "We need the cooperative efforts of scientists in biology, cognitive science and psychology, in addition to computer scientists."

To develop a neurocomputer, Japanese companies are developing four critical elements: an optical mask, a light-emitting diode, a light-receiving photodiode array, and a comparator. In 1988 Mitsubishi Electric developed a simple optical neuron chip using three of these elements to achieve "fuzzy" recognition of visible patterns, which will be useful for future robots and voice recognition systems. Hitachi has combined neural networks and fuzzy logic to develop a new voice recognition technology capable of recognizing twelve voice types. Hamamatsu Photonics and the Institute of Industrial Technology jointly developed an "optical association," an experimental optical neurocomputer with learning and associative memory capabilities that allow it to reproduce images without software programs.

NEC has already announced four neural network systems. One is an alphanumeric character recognition system capable of recognizing almost 100 percent of the characters produced by typewriters and personal computers. Another is a voice recognition system with over 90 percent accuracy in recognizing the numbers 0 to 9 spoken by over one-fifth of all people. This system can be used in future phone translation systems and automated robot systems. The third is a personal neurocomputer based on NEC's PC9800 personal computer, which can recognize 99.95 percent of the characters in seventy-six types and twelve fonts. Featuring a neuro-engine board consisting of four data-flow microprocessors, this machine reduced the software development time required for developing character recognition, voice recognition, and robot control systems by 90 percent. Finally, NEC introduced an optical character recognition (OCR) system based on a neural network accelerator board.

In May 1988 Fujitsu stunned industry observers when it displayed small, neurocomputer-controlled robots at its technology exhibition. In a high-tech game of cops and robbers, Fujitsu developed three "character" robots: Lupin the criminal robot, Sherlock Holmes the detective robot, and a village police robot. These robots have built-in networks of forty-seven neuron model units, using 8-bit microprocessors. They can make 360-degree turns, and they "police" the environment with their twelve optical, ultrasonic, and tactile sensors. In the exhibition, Sherlock Holmes always got his man, the Lupin robot.

MITI's efforts will be matched by neural network computer projects around the world. In 1989 the United States began a two-year, $33 million seed program, which will be followed by an eight-year development project, if all goes well. The European Strategic Programme for Information Technology (ESPRIT) recently announced a $6.4 million seed effort called ANNIE (Applications of Neural Networks for Industry in Europe). The West Germans are conducting a $67 million, ten-year development program, which began in 1987. The Dutch government is planning a $2.5 million seed effort.

Fuzzy Logic Computers

Since 1987 neural networks have been joined by fuzzy logic as the next breakthrough in advanced computing. Invented by Lotfi Zadeh of the University of California at Berkeley in the 1960s, "fuzzy logic" refers to the impreciseness of data in the real world. Whereas existing computers depend on the very precise nature of Boolean logic—where only two truth values are allowed (1 or 0)—fuzzy logic can process "fuzzy" sets of data, such as "small," "near," and "wide." This ability to handle indiscrete pieces of information fits well with the Japanese preference for vagueness and subtlety.

Not surprisingly, Japanese researchers are pioneers in this leading-edge technology. In 1986 Masaki Togai, a Japanese national working toward his U.S. citizenship, codeveloped the world's first fuzzy microprocessors with Hiroyuki Watanabe of AT&T. In 1987, with Rockwell International, Togai developed a second-generation fuzzy microprocessor. With seed funding from Canon and other Japanese companies, Togai founded Togai InfraLogic to develop fuzzy logic computers and silicon chip compiler tools. His company already has a string of firsts: a *kanji* recognition chip for Hitachi capable of 160 million 16-bit operations per second, a 150,000-transistor chip for image processing, a software emulation of fuzzy logic called "AI-Air Conditioner," a digital fuzzy processor, and a fuzzy logic compiler and in-circuit emulator. Togai's fuzzy logic processors may eventually be combined with neural networks to develop powerful machines capable of assigning values to fuzzy data and processing them with their neural networks. His work is being paralleled by Kumamoto University professor Retsusuke Yamakawa, who is developing microprocessors capable of performing fuzzy inferences.

In November 1988 forty-six Japanese companies invested $45.7 million to establish the Fuzzy Engineering Research Laboratory as part of MITI's Basic Technology Research Center. The seven-year International Fuzzy Systems Project involves MITI, STA, Japanese universities, the Japan Fuzzy System Research Association, and foreign companies. Besides promoting an exchange of technical information, the project will conduct both basic and applied research in the fields of fuzzy computing, biomedical fuzzy systems, manufacturing process controls, pattern recognition, robot controllers, and artificial intelligence.

From Bioholonics to Biocomputers

Perhaps Japan's most ambitious program in next-generation computing is its goal of creating biocomputers—computers based on the neural principles of living organisms. If biocomputers capable of mimicking the functions of the brain could be developed, they would be much more sophisticated and flexible than current machines.

Japan's earliest work on biocomputers began in 1982 with the five-year Mizuno Bioholonics Project, which was funded by the Japan Research and Development Corporation (JRDC) as part of STA's ERATO program. The project sought to develop biosensors and biochips capable of sensing, processing, and judging incoming data. The project was divided into two groups. Dr. Denichi Mizuno of Teikyo University headed a team that focused on bioholonics to develop a cancer treatment. Bioholonics is the study of biological processors that form the individual elements of living organisms, such as cells, tissues, and organs. These biological systems were viewed as basic assemblies of synergetic units, or "holons," which could serve as processors in biological parallel processing systems. The advantage of holons over existing computer software is their automatic self-organizing nature, which allows greater system flexibility. If holons could be harnessed, they would provide "intelligence" to conventional AI systems. For example, robots could be told to clean the floor without having to be taught how to sweep;

they would automatically test the floor, generate programs, and execute the required sweeping action.

To pursue this research, senior project associate Professor Hiroshi Shimizu headed a second team that studied neural holons by examining neural network dynamics in the brain. Focusing on the pattern recognition ability of the eye, his team developed a multiprocessor system that simulated a holonic computer, or "holovision" computer, which automatically drew connections between the holons in the system. The holovision computer was able to recognize different shapes, sizes, positions, and orientations of visual images—a key step to image processing and, ultimately, a neural optocomputer.

In 1985 Japanese government ministries and companies became interested in the new field of biocomputing when Dr. Forrest Carter of the U.S. Naval Research Laboratories presented a seminal paper on bioelectronics. Combined with the 1980 Langmuir-Blodgett thin film patent, which opened the door to the mass production of biomaterials, this paper augured well for long-term research in bioelectronic devices, biosensors, and biocomputers. In November 1985 MITI held its own symposium, at which researchers gave a variety of papers on bioelectronics and molecular electronics, including molecular computer design, biological information processing, and vision imaging.

The symposium clearly was influential in Japan. In March 1986 five companies (Kyowa Hakko, Mitsubishi Chemical, Takeda Chemical, Toyo Nenryo, and Toray Industries) formed the Protein Engineering Research Institute to develop biosensors, biochips, biopharmaceuticals, bioreactors, and biofunctional membranes. The $2.5 million project was funded 70 percent by the Key Technology Center and 30 percent by the participating companies. The institute built a $40 million research center in Osaka in 1987 and plans to invest $250 million in protein engineering over the next ten years. Fujitsu, Kirin Brewery, Nihon DEC, and Showa Denko have also joined.

In 1987 MITI began a ten-year project to develop biochips and biocomputers; it is being managed by the quasi-governmental R&D Association for Future Electron Devices. Budgeted at $40 million, the project is exploring the information processing nerve mechanisms of lower animals, such as sea hares and nematodes, and studying their application to biochips and biocomputer architectures. Nine companies are involved in the research: Fujitsu, Hitachi, Matsushita, Mitsubishi Chemical, Mitsubishi Electric, NEC, Sanyo Electric, Sharp, and Sumitomo Electric. The project goal is to develop bioelectronic devices and neural network architectures for a biocomputer capable of mimicking human brain functions, such as pattern recognition, reasoning, and learning. Currently, researchers are developing new molecular electronic structures and biomaterials, and ultrathin-film fabrication techniques.

In corporate laboratories, advanced work on biochips and biocomputers is underway. In 1984 Fujitsu established a five-person biological information lab to study nervous systems, game logic, and parallel processing graphics. In 1985 Mitsubishi developed a biodevice capable of transmitting electrons in a designated direction, opening the way to biocomputing. Sharp is focusing on four areas: molecular recognition, vision data processing molecular devices, biochips, and biocomputers. In 1986 Hitachi formed a bioelectronics group under Masami Naito to develop conducting properties in organic materials using molecular control techniques. And Sony is developing photochromic memories for optical storage media with greater capacities than optical disks.

Overall, more than forty Japanese companies are exploring the new field of bioelectronics. Thus, the Japanese have made a good start on next-generation biocomputing research.

Seventh-Generation Computing and Beyond

In this chapter we have looked at MITI's fifth-generation computer program (symbolic logic, parallel processing, and relational data bases), sixth-generation computers (neural networks, fuzzy logic, and biocomputing), and optocomputing. Yet, there is discussion already about the possibility of a seventh-generation computer.

What features would such a computer have? Stuart M. Dambrot, writing in *Managing Automation* magazine, believes seventh-generation systems will be optical neural computers arranged in holonic-like interacting hierarchies. The computer's logic will be an associative memory system based on the structure of the human cortex, and it will sense the environment using biosensors and "intelligent" biochips. Superconducting wires will form its board-level interconnections and links with electromechanical systems. But most impressive will be the autonomous functions of seventh-generation systems. According to Dambrot:

> Seventh-generation systems will be able to function completely independently, to repair and replicate themselves, and to work more closely with humans. Early seventh-generation robots and mobile systems will sense their power levels, plugging themselves in or exchanging thin-film, lightweight batteries when necessary. Later models such as autonomous, thinking land and space vehicles will incorporate self-contained fusion microreactors.

While still over a decade way, Japanese companies are already working on intelligent robots and the various computer technologies needed to realize the above scenario. It remains to be seen how creatively the Japanese will be able to combine these technologies into new computers.

In this chapter we have reviewed some of the major Japanese projects in next-generation computing. As we enter the 1990s, it is difficult to tell whether these projects will succeed or not, but it is certain that the Japanese will experience numerous failures as they become more daring and creative in their research. Nevertheless, they are sure to rapidly commercialize whatever dramatic breakthroughs come their way as their advanced computer projects come to fruition. That is the genius and the challenge of Japanese industry—to rebound from failure and learn from it.

12

Visions of Superconductors

> American companies, by and large, have taken a conservative, wait-and-see attitude; they may have already begun to fall behind.
>
> —U.S. Office of Technology Assessment

ONLY days after J. Georg Bednorz and K. Alex Mueller of IBM Zurich announced their discovery of high-temperature superconductivity in November 1986, Professor Shoji Tanaka of Tokyo University happened to show their paper to one of his graduate students. He was skeptical. If it was true, IBM's discovery of new ceramic oxides was epoch-making. Superconductivity was discovered in 1911, and the goal of developing high-temperature superconductivity—materials with no resistance to electrical current at temperatures higher than 0 Kelvin, or absolute zero—had remained an elusive dream for seventy-five years. Niobium-based superconductors were used in medical scanning equipment, but they peaked at 23 Kelvin.

To Professor Tanaka's amazement, his student duplicated IBM's findings, achieving superconductivity at 35 Kelvin with a mixture of rare earth metals, including lanthanum, barium, and strontium. Immediately, Professor Tanaka reported his findings to the scientific community. It was no longer a quixotic search; IBM's findings were real. The race for high-temperature superconductivity was on!

Superconductor Fever

Around the world, thousands of scientists scrambled to investigate the new superconducting materials (see Table 12–1). Japan's Science and Technology Agency (STA) immediately held a series of meetings around the country and formed an advisory group, the New Superconductivity Materials Forum, to explore the implications of the new ceramic oxide superconductors. Bell Laboratories achieved superconductivity at 36 Kelvin using a new ceramic oxide compound of lanthanum, strontium, and copper. Professor Kazuo Fueki of Tokyo University reached 37 Kelvin using a similar material. At the Tsukuba Science City, researchers at MITI's Electrotechnical Laboratory raised the standard to 46 Kelvin with a new mixture of lanthanum, strontium, and copper oxides. These startling results sent hundreds of Japanese companies into action.

In February 1987 Professor Ching-Wu "Paul" Chu of Houston University shook the scientific world with a spectacular breakthrough of vast commercial potential. By mixing the rare earth yttrium, instead of lanthanum, with barium, copper, and

TABLE 12–1
Breakthroughs in High-Temperature Superconductivity

Date	Group	Kelvin	Materials
1911	Heike Kamerlingh Onnes (Holland)	4	Mercury
1973	John Cooper	23	NiGe
November 1986	IBM Zurich	30	LaBaCuO
December 1986	Tokyo University (Shoji Tanaka)	35	LaBaSrO
December 1986	Bell Laboratories	36	LaSrCuO
December 1986	Tokyo University (Kazuo Fueki)	37	LaSrCuO
December 1986	Houston University (Paul Chu)	39	LaSrCuO
January 1987	Okazaki National Research Institute	43	LaSrCuO
January 1987	MITI Electrotechnical Lab	46	LaSrCuO
February 1987	Houston University (Paul Chu)	98	YBaCuO
February 1987	Chinese Academy of Sciences	100	YBaCuO
March 1987	Japan National Research Institute for Metals (NRIM)	123	YBaCuO
January 1988	NRIM	106	BiSrCaCuO

NOTE: Lanthanum (La), Barium (Ba), Copper (Cu), Oxygen (O), Yttrium (Y), Strontium (Sr), Bismith (Bi), Niobium (Ni), Germanium (Ge), Calcium (Ca)

oxygen, he achieved superconductivity at 98 Kelvin. He had broken the magic barrier—77 Kelvin—above which scientists could achieve superconductivity by using inexpensive liquid nitrogen that costs only 22¢ per gallon. Now it was possible to develop commercial-grade superconductors without having to use liquid helium costing $11 per gallon.

The commercial prospects of Professor Chu's discovery floored the Japanese. Four days after the "Houston shock," MITI responded by announcing plans to form a superconductivity research consortium of Japanese companies, universities, and government laboratories. MITI's founding group included Mitsubishi Electric, NTT, Hitachi, Matsushita Electric, Kyocera, and Tokyo Electric Power. Experts in the United States and Europe were worried, and rightfully so. Japan had decades of experience in niobium-based superconductivity, ceramics, and superconducting trains. Now it had pulled together its top researchers to share information and chart a national strategy for the new superconductors. Would Japan dominate yet another emerging industry?

At Tokyo University, Professor Shinichi Uchida—a protégé of Professor Tanaka—quickly formed a thirteen-person research team that worked around the clock, seven days a week, to verify the Houston discovery. The researchers slept on bunk beds and cooked meals in a kitchenette in temporary quarters so that they could continue firing ceramic pellets without interruptions in the laboratory's kiln. They tried hundreds of combinations and were sidetracked into pursuing the wrong element—ytterbium—by an erroneous news report from China. Three weeks later, in early March 1987, they succeeded in duplicating Professor Chu's discovery. Immediately, Professor Uchida's team was flooded with visitors from Fujitsu, Hitachi, Sumitomo Electric, Toshiba, and other Japanese companies.

Their news was overshadowed by a big breakthrough across town at the Tsukuba laboratory of the National Research Institute of Metals (NRIM). A group led by Kazuma Togano claimed superconductivity at 123 Kelvin using a totally new material, bismuth. Coming on the heels of Tokyo University's announcement, the Tsukuba breakthrough intensified "superconductor fever" in Japan. Throughout Tokyo, companies rushed to develop new superconducting materials and file patents in hopes of getting into this emerging technology. MITI's new materials program—a consortium of metal, ceramics, and composites makers—mobilized its members.

The superconductor boom swept Japan with gale force. Nonferrous metal firms rushed into the rare earth market to meet the rapidly growing demand for research materials. Major players such as Asahi Chemical, Kawasaki Steel, Mitsubishi Chemical, Nippon Steel, Shinetsu Chemical, Sumitomo Metal Mining, and Toyo Soda were swamped with orders. The Japan Society of Newer Metals reported a huge jump in demand for rare earth metals. MITI began surveying worldwide sources of rare earth metals and entered into an agreement with the Chinese government to explore China's western regions for the new rare earth materials.

Announcements from Japan came fast and furiously. In 1987 Toshiba developed a new ceramic material, superconductive at 93.7 Kelvin, that could be shaped into wire rods. NTT Corporation came up with an unusual holmium-barium-copper oxide thin film for future computer circuits. Hitachi announced high-speed infrared detectors using a superconducting film, an optical switch, and a thallium-based ceramic superconductor. NRIM created a variety of superconducting wires. Fujitsu announced an yttrium-barium-copper oxide circuit on an aluminum substrate with zero resistance at a phenomenally high 159 Kelvin. Mitsubishi Metal sampled ceramics-based sputtering targets for superconducting circuits. NEC displayed ultrafast circuits—a superconducting quantum interference device (SQUID) for medical uses and a Josephson circuit for computers. Matsushita Electric found a way to deposit four layers of thin superconducting films. Sharp introduced superconductive magnetic sensors.

The biggest announcement came from Sumitomo Electric, which shocked industry watchers by filing over 700 patents in 1987 covering everything from superconducting thin films to wires and rods. Sumitomo has already run into a patent conflict with two MITI professors at American Superconductor because of differences in patent laws. In the United States, patent rights go to the first to conceive, rather than the first to file, as in Japan; the stage is set for future patent fights. In Europe, the British Technology Group reported that Japanese companies are filing numerous patent applications covering basic technology and processing methods. The battle lines are being drawn.

By early 1988 Japanese companies were already sampling prototype superconductive magnetic sensors, for which they have plans for full-scale commercialization in the early 1990s. They are well positioned to parlay their strengths in consumer electronics, thin wire formation, ceramics, micromanufacturing, semiconductor equipment, magnetic levitation trains, photovoltaics, and batteries. Moreover, they could take these new technologies to market quickly because of their strength in applications engineering and manufacturing know-how.

In a report to the U.S. Congress, the Office of Technology Assessment (OTA) warned:

> Corporate executives in Japan . . . see high-temperature as a major new opportunity—one that could set the pattern of international competition for the 21st century. Japanese companies have made substantial commitments of people and funds, pursuing research and applications-related work in parallel. Firms in more lines of business are at work than in the United States. Steel companies and glassmakers, as well as chemical producers and electronics manufacturers, are seeking new businesses, ways to diversify. Japanese managers see in high-temperature superconductivity a road to continued expansion and exporting, and are willing to take the risks that follow from such a view.

OTA found that thirty-eight major Japanese companies had assigned 900 researchers to superconductor research in 1987, spending about $90 million. By contrast, U.S. companies had assigned 625 researchers and had spent $97 million. Overall, OTA estimated total U.S. superconductor spending in 1988 would reach about $250 million, versus $160 million for Japan. But the report warned that these figures underestimated Japan's commercial efforts. OTA estimated that Japan is spending half of its R&D funds on commercializing superconductors, while the United States spends 70 percent on defense-related research.

Rapid commercialization is clearly what the Japanese have in mind. In mid-1987 the *Japan Economic Journal* found that half of the Japanese researchers surveyed thought superconducting materials would be commercially available by the early 1990s. Local market researchers forecast that the Japanese superconductor market will reach $8.3–25 billion by the year 2000. If room-temperature superconductors are developed, they believe this figure could reach $83 billion. Japanese companies are investing heavily to establish themselves in this highly competitive arena.

The Bureaucrats Vie for Control

Japanese companies and university researchers are not the only players trying to get a foot in the door. Since Professor Tanaka's explosive finding, government bureaucrats have been fighting turf battles for funding and jurisdiction. Superconductivity is one of the hottest topics in Tokyo. Based on information from the U.S. embassy in Tokyo, OTA reported: "Superconductor fever has swept through Japan's government too, with ministries vying with one another for the lead in policy. The picture has now stabilized, but 1987 saw many actors seeking center stage—and few signs of the coordinated, monolithic policy machine that some Americans still think of as Japan, Inc."

Japanese government funding of superconductor research is as much of a hodgepodge as U.S. government funding (see Table 12–2). Funding allocations and priorities are based on political jurisdictions, as well as pure clout with the Ministry of Finance (MOF). STA handles science policies among public, private, and academic researchers. MITI focuses on the commercial application of technology. The Ministry of Education (MOE) funds basic research in the universities. The Ministry of

TABLE 12–2
Japanese Government Superconductivity Funding (fiscal 1988)

Ministry	Research Focus	Amount ($ millions)
International Trade and Industry (MITI)	Room-temperature materials Innovative processing Supercomputing devices Power generation/storage International research center	$26.2
Science and Technology Agency (STA)	New superconducting materials Traditional superconductors Multicore Project	24.6
Transport (MOT)	Linear motorcar development	4.6
Education (MOE)	Research facilities and grants	17.6
Posts and Telecommunications (MPT)	Electric telecommunications	0.5
Total		$73.5

SOURCE: International Superconductivity Technology Center (ISTEC).

Posts and Telecommunications (MPT) is exploring electrical transmission and storage technologies.

In April 1987 MITI's Agency for Industrial Science and Technology (AIST) took the initiative by holding an industrial superconductor conference with Professor Tanaka. MITI's Future Industry Basic Technology R&D Project annually spends around $6 million for research into superconductivity, while its Supercomputer Project budgets around $4 million for ultrafast Josephson junction devices. In July 1987 MITI launched a study to explore the feasibility of developing superconducting materials aboard space stations during the 1990s. In May 1988 MITI added high-temperature superconductors to its Next-Generation Basic Industrial Technology program. This ten-year program will focus on magnetic levitation trains and electric cars, machinery, electronics, and electric power industries. MITI's Moonlight Project, which focuses on energy conservation, will spend around $60 million between 1988 and 1995 on superconducting electrical generators and wires.

In April 1987 STA announced plans for a new superconductor program involving private industry, universities, and government. It incorporated the New Superconductivity Material Research Corporation and created the separate Superconductive Materials Research Center. In 1988 STA initiated its superconductor program, the Multicore Project, which is budgeted at $160 million, $16.4 million of which was allocated for 1988. The project involves 150 scientists focusing on fourteen research topics (see Table 12–3). STA is working with NRIM, the National Inorganic Materials Research Institute, and MIT's Magnet Center to develop an information exchange system and a data base on superconductivity.

MOE is responsible for setting educational policies and granting research subsidies to Japanese national universities. In 1984 MOE qualified new superconducting materials research for financial support under its special project research (*tokutei kenkyu*) class of grants-in-aid programs. Between 1984 and 1986, MOE granted $2.65 million to fifty-one superconductor projects. Major grants of between $41,700 and $417,000 are awarded to groups of five to ten scientists. In June 1987 MOE established a $1.4 billion science fund to seed superconductivity research

TABLE 12–3
STA's Superconductor Multicore Project
(fiscal 1988)

Research Core	Amount ($ millions)
Superconductivity theory	0.04
Research data base	0.39
Process exploration	0.75
Purification processes	0.00
Thin-film processing	1.47
Single crystal processing	0.22
Lithography	0.19
Conductor processing	0.00
Space environment processing	0.00
High-field properties	7.23
Crystal structures	2.80
Chemical compositions	0.68
Neutron irradiation	2.67
Physical and analysis	0.00
Total	16.44

SOURCE: U.S. State Department.

at a dozen universities. Top funding recipients include professors Shoji Tanaka, Fumihiko Takei, and Sadao Nakajima (Tokyo University), Yoshio Muto (Tohoku University), Ikuji Tsujikawa (Kyoto University), and Ko Yasukochi (Nihon University).

The Ministry of Transport (MOT) has been the main driver behind Japan Railway's linear motorcar program, while MPT has established its own superconductivity research project for superconducting wires and transmission lines.

In January 1988 MITI again took the lead by banding together with eighty-eight Japanese companies to form the International Superconductivity Technology Center (ISTEC) as a forum for basic research and industry seminars. Professor Tanaka heads research at ISTEC, which is a nonprofit corporation with over 100 members. As of early 1989, forty-five companies paid the full membership fee of $833,000, plus $100,000 in annual fees, which entitle them to send their researchers to the ISTEC laboratory, as well as to a share in the intellectual property created there. The other fifty-seven companies are associate members who are allowed to participate in selected programs and workshops. MITI provides about $3.75 million annually in consignment funding (*itakuhi kenkyu*). Besides its impressive roster of companies, ISTEC managers include Gaishi Hiraiwa, vice chairman of the powerful Keidanren (Federation of Economic Organizations); Katsuhige Mita, president of Hitachi; Tetsuro Kawakami, president of Sumitomo Electric; and Takashi Akutsu, a MITI official formerly assigned to the San Francisco consulate. Thus, ISTEC is linked to Japan's centers of economic and political power.

Although ISTEC has opened its doors, foreign companies have been reluctant to join because of the high membership fees and their fear of losing technology to Japanese competitors. Taking the initiative, ISTEC ran a full-page ad in the *Wall Street Journal* challenging U.S. companies to join. But foreign companies held back until ISTEC's intellectual property and patent rights guidelines were issued in late 1988. Under the proposed guidelines, ISTEC will have rights to 50 percent of all patents for anything developed in its laboratory, with the remaining 50 percent divided among the codevelopers of the patented item.

As of mid-1989, there were five foreign companies in ISTEC: British Telecom, DuPont Japan, IBM Japan, Intermagnetics General, and Rockwell International. Most other foreign companies are still skeptical about ISTEC's claim that it is an international body because of its tight circle of players. ISTEC has seven research divisions led by top researchers from MITI, Tokyo University, and NEC, illustrating Japan's strong industry-university-government ties in this emerging field (see Table 12–4). Six of the divisions are located in Tokyo; the Nagoya Research Division is located at the Japan Fine Ceramics Center near MITI's regional testing laboratory in Nagoya City.

ISTEC will be the first genuine test of Japan's claim that it is opening up to foreign companies. Although the Fifth Generation Computer Project holds international conferences and has opened its laboratory to foreign visitors, foreign staff researchers have not been actively recruited, nor have major foreign companies like IBM Japan or Digital Equipment been allowed to join. Although ISTEC has formally opened its membership, it did so in response to pressure from the U.S. and European governments. The informal "old boy" network in Japan will be tough to crack for foreign companies, even for those that hire Japanese nationals. Leading-edge research in

TABLE 12–4
Research Divisions of the International Superconductivity Research Laboratory

Division	Research Focus	Leader
I	Fundamental research	Dr. Koshizuka, MITI Electrotechnical Laboratory
II	High-temperature superconducting oxides	Dr. Yamanouchi, Tokyo University
III	Organic superconductors	To be determined
IV	Chemical processing	Dr. Shiobara, MIT graduate
V	Physical processing	Dr. Morishita, NEC Corporation
VI	Data bases	Dr. Ishiguro, MITI
VII	Ceramic processing (Nagoya Division)	Dr. Hirobayashi, Tokyo University

SOURCE: International Superconductivity Technology Center (ISTEC).

Japan is usually discussed among companies after hours in bars and restaurants. Unless foreign companies have access to this informal network, which takes years to develop, they will be locked out of the key research results. Moreover, if ISTEC's Japanese members make stunning breakthroughs in superconductivity, Japan will be absolved of its responsibility to foreign governments by having opened its doors earlier.

Whatever the outcome, ISTEC is a major force to be reckoned with in the superconductor world. Whether it will go the way of "techno-nationalism" or "techno-globalism" depends in large part on the reaction in Silicon Valley, Washington, D.C., Paris, Bonn, and London.

The 300-mph Superconducting Train

The sleek, cigar-shaped linear motorcar skims over the elevated concrete guideway at 320 miles per hour, making a hissing sound as it zips past the palm trees and rice paddies bordering the oval test track. In the distance, the two-car "maglev" (for magnetic levitation) is no more than a speck as it approaches from down the straightaway of the three-mile-long track, but within seconds it whooshes past the observation platform in a blur, only thirty feet away. Then, as quickly as it arrived, the maglev is gone, leaving the startled observer gripping the guardrail. Welcome to Hyuga—test site for Japan Railway's bold experiment in superconducting linear motorcars, the next generation of ultrafast bullet trains in Japan.

Since the 1960s, Japan has experimented with maglev technology using linear-motor propulsion and no-contact suspension. In 1972, under the leadership of Dr. Yoshihiro Kyotani, its former director of technical development, Japan National Railway (JNR) developed a linear synchronous motor (LSM) propulsion system that is used in its linear motorcars. Since opening its Hyuga test track in 1977, Japan Railway (privatized in 1987) has made great strides in maglev technology. In 1978 the MLU-500 single-car test vehicle reached a top speed of 187 miles per hour, then surpassed that record in December 1979 when it topped 320 miles per hour using superconducting magnets.

In 1981 a redesigned two-car vehicle, the MLU-001, was tested on a new U-shaped track; it reached 155 miles per hour. The MLU-001 featured two superconducting coils and helium indirect cooling technology. In February 1987 it achieved 252 miles per hour. Since March 1987 Japan Railway has tested the more powerful MLU-002, which features 17-ton cars seating up to forty-four people, safety monitoring cameras, and LCD TVs on the seat backs so passengers can see the oncoming track. This system is being developed by Japan Railway's Technical Research Institute in Tokyo, headed by Dr. Masanori Oeki. Japan Railway spends about $15.4 million annually on superconducting linear motorcars. In 1987 sixty member companies of the Future Technology Research Group forecasted that superconducting linear motorcars controlled by supercomputer systems would be deployed by the year 2005.

Linear motorcars are being developed because they would relieve some of the pressure on the Tokaido Shinkansen, Japan's bullet train, whose lines have begun to show serious signs of wear from the constant pounding of 130-mile-per-hour trains. Each night the tracks must be carefully realigned by work crews. Worn-out rails must be replaced and slipping concrete stakes shored up. To overcome these hazards, Dr. Kyotani devised dynamic maglev technology using repulsion (as opposed to magnetic attraction, which is used in West Germany) to lift the train cars from the tracks.

Based on the theoretical work of G. Powell and J. R. Danby of Brookhaven National Laboratory, the Japan Railway technology achieves magnetic levitation by on-board superconducting magnets that repulse conductive coils embedded in a U-shaped guideway. Liquid-helium refrigeration units on the train supercool the magnets to 4.2 Kelvin, maintaining current within the magnets even when the power is cut off. When the train passes over the guideway coils, a strong magnetic field is generated, repelling the train about four inches off the track. To propel the maglev cars forward, superconducting magnets with alternating polarities (positive and negative) simultaneously push and pull the magnets on the cars. Each propulsion magnet pushes a magnet on the car, reverses polarity and pulls the next magnet, then reverses polarity again and pushes the next magnet. The linear motorcar can be accelerated by speeding up the current in the superconducting magnets in the guideways.

Dr. Kyotani—president of the consulting firm Technova and the "Werner von Braun" of Japan's linear motorcar program—is an ardent proponent of maglev technology. Currently, Japan Railway and MOT are reviewing seven proposed routes for the linear motorcar. According to Dr. Kyotani,

> The most serious proposal is the Tokyo-Osaka route that runs 500 kilometers through the city of Kofu, through the mountains, and past Nagoya. MOT will announce its decision on this proposal by late 1989 or early 1990. This route would cost at least 3 to 4 trillion yen [$25–33 billion], an average cost of 6 to 8 billion yen per kilometer [$80–110 million per mile]. The land in the Tokyo area alone would cost at least one trillion yen [$8.3 billion]. To reduce land acquisition and construction costs, the project would have to be built quickly—in four to five years. If this route is built, it will only take 20 minutes to go from Tokyo to Kofu, 20 minutes more to Nagoya, and another 20 minutes to Osaka, instead of the present 3 hours from Tokyo to Osaka. Office workers could live in Kofu and commute to Shinjuku station in Tokyo.

Shoji Sumita, president of East Japan Railway, challenges Dr. Kyotani's cost estimates. He believes that the Tokyo–Osaka route would cost more than 10 trillion yen ($83 billion), owing to land speculation.

Currently, three Tokyo–Osaka plans are being considered. In the first plan, the Japan Transportation Engineering Company (JTEC) would attract private Japanese investors who would otherwise invest their money overseas. The second plan is a national project in which Japan Railway would pay for two thirds of the cost, and the Japanese government one-third. The third plan is for 100 percent Japan Railway ownership, with the company putting some of its profits into a "linear construction fund."

Linear motorcars are a hot topic in Tokyo because of their potential for creating thousands of construction jobs, political patronage, and huge markets at the stroke of a pen. Contracts for new computer and telecommunications systems, travel and retail services, real estate development, and advertising would run into billions of dollars. The economic benefits of a Tokyo–Osaka linear motorcar project would be enormous. Besides annual construction costs of $3–4 billion, Mitsubishi Research Institute estimates this project would generate $87 billion in new domestic business and $62 million in additional tax revenues. It would open up new housing areas, reduce housing pressures in Tokyo, and trigger urban development in Kofu, Nagoya, and Osaka. The distance between Tokyo and Osaka would be compressed, making one-day business and vacation trips easy and economical.

Critics argue that a Tokyo–Osaka route would accelerate the rural exodus, lead to uncontrolled land speculation and kickback schemes (as happened with the Shinkansen in the 1970s), and create a congested Tokyo–Nagoya–Osaka "supercity." Given Japan's pervasive "money politics" and loose insider trading laws, not to mention the Shinkansen experience, their criticisms are probably valid.

Moreover, JR Tokai, the regional Japan Railway company responsible for operations in the Tokyo area, estimates the linear motorcar would siphon away up to 74 percent of the Shinkansen passengers, turning the Shinkansen gold mine into a money loser. According to Kazumi Sotoyama of Japan Railway's public relations department, the Tokyo–Osaka linear motorcar would bring in $36.7 billion in revenues over a thirty-year period, leaving the operation $11.7 billion in the black. But this would be offset by a loss of $20.8 billion by the Shinkansen. The net loss to Japan Railways would be $9.2 billion. But Dr. Kyotani suggests another way of looking at the situation: "One way to reduce this loss is to turn the Tokaido Shinkansen into a local commuter train by charging lower fares. Already, many people commute by Shinkansen because of Tokyo's high housing costs. Their monthly mortgage payments are reduced by more than the cost of the bullet train tickets."

In January 1988 Japan Railway consolidated the linear motorcar research of the seven regional railway companies into its central laboratory. Each company will be levied 0.35 percent of their revenues to finance research on new superconducting magnets for Japan Railway's planned MLU-003 linear motorcar. In preparation for MOT's decision, major construction, real estate, and electrical machinery companies are gearing up their linear motorcar planning divisions. Hitachi, Mitsubishi Electric, and Toshiba, for example, have joined hands to pursue independent superconducting magnet research paralleling Japan Railway's research efforts. They see big business coming from the linear motorcar project if it is approved by MOT.

The Tokyo–Osaka route is being hotly debated within government circles and the media. Although MOF opposes government funding for the scheme, MOT and MOC (the Ministry of Construction) strongly support the project. In spite of Japan Railway's consolidation of research, a power struggle has erupted among its seven companies for control over funding and project management. The regional companies fear that JR Tokai will grab power and delay prospects for linear motorcar or Shinkansen routes elsewhere in the country. As these machinations indicate, the superconducting linear motorcar is no longer just Dr. Kyotani's visionary dream, but a heated political debate that will deeply influence the direction of Japan's superconductivity research, as well as its land and rail development.

From Medical Scanners to Ultrafast Ships

Linear motorcars may be the hottest topic in Tokyo, but they are not the only commercial prospect being targeted by Japanese companies. In a poll taken by the Nikkei Industry Research Institute, industry experts forecast that the Japanese market for superconducting products will reach $12.4 billion within ten years. If superconducting materials capable of operating at room temperature are developed, the market could reach $85 billion or more. As shown in Table 12-5, superconductors are certain to be used (if they are not already being used) in seven

TABLE 12-5
Superconductor Market Forecast by Japanese Experts

Applications	Probability in 10 Years (%)	Predicted Value ($ millions)
Magnetic levitation railway	100.0	4,000.00
NMR-CT	100.0	303.00
Free electron laser	100.0	120.00
SQUID	100.0	58.40
Compact cyclotron	100.0	43.20
Large-scale particle accelerator	100.0	17.60
Thin-film target materials	100.0	8.00
Zero-resistance circuit boards	95.0	801.00
Electromagnetic launching equipment	95.0	7.60
Compact SOR	90.0	509.00
Magnetic shields	90.0	72.09
Oscilloscopes	90.0	14.40
Ultraviolet sensors	80.0	6.40
Magnetic separation equipment	75.0	120.00
Large-scale electric power storage	70.0	2,168.00
Standard voltmeters	65.0	0.70
Electric power generators	56.0	35.60
Zero-resistance LSI circuitry	55.0	507.80
Motors	52.5	59.20
Electric power lines	52.5	21.00
Supercomputers	50.0	1,385.00
Power generators	50.0	20.00
Superconducting LSIs	45.0	254.56
Magnetic propelled ships	30.0	1,200.00
Magnetic energy storage	25.0	6.20
Transformers	25.0	6.00
Electric automobiles	20.0	640.00
Electric ships	10.0	8.00
Modulators, demodulators	0.0	0.00
Electron scanning microscopes	0.0	0.00
Home power storage	0.0	0.00
Low-loss communications cables	0.0	0.00
Total		12,392.75

NOTE: $1 = 125 yen
NMR-CT: nuclear magnetic resonance computerized tomography
SQUID: superconductive quantum interference device
SOR: synchroton orbit radiation
LSI: large-scale integration
SOURCE: Nikkei Industry Research Institute.

of the many applications listed: linear motorcars using magnetic levitation, nuclear magnetic resonance (NMR) medical scanners, free electron lasers, SQUIDs (superconducting quantum interference devices for medical sensors), compact cyclotrons for scientific research, large-scale particle accelerators (like the proposed "supercollider" in the United States), and thin-film target materials for electronic systems. By late 1987, according to the Cambridge Report on Superconductivity, "an avalanche of superconducting product and new announcements is pouring forth. . . . Some developments are in early stages, while others are said to be ready for immediate or very near-term commercialization. Nearly all are from Japan."

The medical field is especially hot. Because of Japan's aging population, the demand for better medical treatment and monitoring systems is growing. Currently, low-temperature superconducting magnets are used in magnetic resonance imaging (MRI) machines that provide detailed images of the chemical composition of the human body. The procedure is noninvasive and does not require X-rays. MRI machines now cost around $1.5 million; high-temperature superconductors would make it possible to design smaller, lighter, and less costly machines. Japanese companies are working on portable MRI scanners to reduce systems costs even further for medical clinics and hospitals. These scanners will employ SQUIDs that are extremely sensitive to faint electromagnetic signals, such as the neural impulses in the brain and heart.

Particle accelerators using low-temperature superconducting magnets are also grabbing attention in Japan because of their use in basic research. In October 1986 the TRISTAN (Transposable Ring-Intersecting Storage Accelerator in Nippon) ring was completed by the Ministry of Education's High-Energy Physics Research Laboratory at a cost of $620 million to conduct research on the sixth quark, the key to understanding the nature of matter and the origin of the universe. Circular accelerators, such as TRISTAN or the proposed supercollider in Texas, send two beams of particles in opposite directions, causing them to smash together. Scientists then study these collisions. Low-temperature superconducting magnets save millions of dollars in electricity costs over electromagnets; high-temperature superconducting magnets would reduce electricity and refrigeration costs even more.

Beyond the sure bets, Japanese companies are focusing on the biggest potential markets. Large-scale electrical power storage ($2.2 billion), supercomputers ($1.4 billion), magnetically propelled ships ($1.2 billion), computer circuit boards ($801 million), and electric cars ($640 million) have captured much attention in Japanese corporate laboratories. These markets, however, are not likely to gain momentum for another fifteen to twenty years, so the government ministries will probably organize a variety of joint research projects to assist corporate researchers.

Superconducting Chips and Computers

In global economic terms, the real battle for supremacy in superconductivity lies in electronics, which Japanese experts believe will be a $3.5 billion market in ten years. Superconducting electronic devices and interconnections not only would increase computing speeds but would also reduce electrical costs and system overheating. The market for superconducting interconnecting wires for zero-resistance computer circuit boards, for example, is expected to reach $800 million by 1998. Fujitsu, Hitachi, NEC, and other companies are developing new types of thin films using rare earth elements, such as yttrium, holmium, and bismuth, to develop chip interconnects. Sumitomo Electric recently developed a notable holmium-compound superconducting thin film that can carry 3.5 million amperes per square centimeter (high current densities are required for computers) at 77 Kelvin. NTT Corporation has a holmium thin film that carries 2.6 million amperes per square centimeter. Around the world, companies are racing to develop new manufacturing methods for creating these circuit board interconnections.

Superconducting LSI (large-scale integration) chips are also attracting attention. Since 1981, MITI's Supercomputer Project has been developing Josephson junction devices, which are superconducting electronic circuits with switching times as low as six picoseconds (one trillionth of a second), or ten times faster than existing devices. After IBM dropped out of the field in 1983, Japanese companies made rapid progress. In 1987 NEC applied for 100 patents for its Josephson junction devices that operate at liquid nitrogen temperatures. In 1988 Fujitsu announced a niobium-based 4-bit microprocessor on Josephson junctions. Its superfast processing speed (770 megahertz) is twenty-five times faster than silicon-based microprocessors and ten times faster than gallium arsenide–based devices. Hitachi has developed a CAD system that automates the wiring of Josephson junction chips. The CAD system can design a 100-gate Josephson circuit in one-tenth the time taken by current approaches. Hitachi has developed superconducting thin films for optical switching devices and mass production technology for SQUID sensors. Meanwhile, researchers at STA have developed a new Josephson junction device using a new technique to deposit yttrium onto a titanium oxide strontium substrate. Japanese companies hope to make new supercomputer breakthroughs using these new ceramic superconductors.

Superconductivity is also useful for manufacturing ultra-dense silicon chips. One emerging field is synchrotron orbital radiation (SOR), which has been developed at Brookhaven National Laboratories in the United States for many years. An SOR ring is an oval- or doughnut-shaped machine that uses powerful magnets to accelerate electrons to almost the speed of light. It bends the orbit of the light by applying strong magnetic fields to generate high-energy X-ray beams from the accelerated electrons. SOR beams are one million times more intense than the photolithographic light used in current chip-making equipment. They can draw extremely fine lines on silicon, down to 0.1 microns (one ten-millionth of a meter). For comparison's sake, one human hair is about 75 microns across.

Since 1982, MOE's High Energy Physics Laboratory in the Tsukuba Science City has operated a 28-meter–diameter SOR ring to conduct basic research. Two large superconducting SOR rings are being planned in Japan: the $833 million Kansai SOR

Development Center in the West Harima Technopolis near Kobe, and the $66 million Synchrotron Applications Research Center in the Hiroshima Technopolis. Currently, nineteen SOR rings are under construction in Japan.

Conventional SOR rings are too big across, however, to fit into chip-making plants. To make them commercially feasible, Japanese companies are busy developing compact SOR rings less than one meter in diameter. These compact SOR rings will be used to mass-produce ultradense memory chips, beginning with 64-megabit prototypes in the early 1990s. Initially, compact SOR rings will use electromagnets; later, they will be switched to superconducting magnets. Sumitomo Heavy Industries plans to commercialize its half-meter–diameter SOR ring in the early 1990s.

In 1986 MITI's Key Technology Center and thirteen Japanese chip makers formed Sortech, a $100 million joint venture company, to develop compact SOR equipment in time for mass-producing 64-megabit memories in 1995. The members are a "who's who" of Japan's semiconductor industry: Canon, Fujitsu, Hitachi, Matsushita Electric, Mitsubishi Electric, NEC, Nikon, Oki Electric, Sanyo, Sharp, Sony, Sumitomo Electric, and Toshiba. Developing compact superconducting SOR rings is the key to dominating the future of high-density memory chips and, ultimately, the computer and consumer markets.

Building Superconducting Energy Systems

Energy has always been a major concern for Japan, which imports more than 80 percent of its oil from the Middle East. After the oil shocks of 1973 and 1979, the Japanese government implemented a nationwide program of oil stockpiling, energy conservation, nuclear power plant construction, and alternative energy development. Now superconductivity offers the prospect of large-scale systems that would enable Japan to generate electricity and store it in underground systems for long periods of time. According to the Nikkei Industrial Research Institute, large-scale superconducting electrical power storage will be a

$2.2 billion market in ten years, while electrical power generators and power lines will reach a modest $83 million.

In the fall of 1987 thirteen Japanese companies agreed to fund a ten-year, $200 million project to build a 200-megawatt electric generator using low-temperature superconductor technology. The Superconductivity Generator Materials and Research Association (Super-GM) is headed by three of Japan's electric utilities—Chubu Electric Power, Kansai Electric Power, and Tokyo Electric Power—and includes the Fine Ceramic Center and the Japan Electric Power Research Institute. Headed by Kiyoji Mori, president of Kansai Electric Power, the group will initially build a test 70-megawatt generator. Maekawa Corporation and Ishikawajima-Harima are conducting research on cooling systems. Electrical equipment companies such as Hitachi, Mitsubishi, Fujikura, Furukawa, Toshiba, and Sumitomo are developing power generators and transmission equipment. By 1996 the group hopes to develop a prototype superconducting generator that can be commercialized by the member companies. The Japanese government is providing additional research funds by levying taxes on electricity sold by the nine Japanese regional utilities.

The strategic focus of Japan's research into superconducting energy systems is electrical storage, not generators. Why is this? What are the advantages of stockpiling electricity instead of relatively inexpensive oil? Probably the key concern is Japan's dependency on oil-producing nations and oil shipping lanes, which are extremely vulnerable to political pressures and military conflicts. Superconducting storage systems would allow Japan to rely more heavily on hydroelectrical, geothermal, solar, biomass, wind, and alternative forms of energy, without worrying about blackouts and brownouts. Electrical power could be stored indefinitely in large underground "batteries," which could provide peak capacity during demand surges. Professor Shoji Tanaka of Tokyo University estimates that if the efficiency of superconductors were increased 30 to 40 percent, either oil-fueled or nuclear-powered generation could be eliminated in Japan.

Another benefit of superconducting storage systems would

be less environmental damage from oil spills, which have occurred several times along Japanese coastal areas, and from air pollution generated by oil-burning systems. Underground storage systems could be located near urban areas where demand is highest. Unlike oil storage tanks, superconducting storage systems would be "clean"; they would neither explode by accident nor pollute the environment. Moreover, implementing this technology would create thousands of badly needed construction jobs, new underground construction technologies for Tokyo's "geopolis" (underground city) program, and new superconductor supply industries. For an energy-poor, land-scarce nation like Japan, superconducting storage systems make political, economic, and environmental sense.

To meet this challenge, MITI has initiated a two-stage program to develop superconductors for electrical power storage and distribution. In phase 1 (1988–90) MITI is surveying superconducting magnetic storage research efforts around the world and setting development priorities. In phase II (1990–98) a joint industry-university-government project will be formed to develop superconducting generators and storage systems. One such joint project is being pursued by Hitachi and Chubu Electric Power, which are spending about $7.5 million each to develop a prototype superconducting magnetic power energy storage (SMES) system by early 1990.

Parallel to MITI's program is the Mukaibo Project, which was organized in February 1988. Dr. Takashi Mukaibo, acting chairman of Japan's Atomic Energy Commission, formed a study group to develop a plan for an experimental superconducting plant. His plan calls for building a large niobium-titanium alloy coil, cooled with helium, inside a rock formation, possibly near Mount Tsukuba, northeast of Tokyo. Eventually, Mukaibo envisions building coils with a storage capacity of five million kilowatt hours and an output of one million kilowatts. This system could store excess electrical power produced at night and release it during the day at peak demand. His project is being assisted by the New Energy Development Organization and the Engineering Advancement Association of Japan, which have conducted research on superconductivity since 1968.

Superconductus Nipponica?

Thus, Japan is pushing into creative superconductor research with lightning speed. In this chapter we have reviewed some of Japan's major programs and their ramifications over the next ten years. While it is impossible to review all of the programs proliferating among Japanese companies, the general contours of Japan's superconductor strategy are becoming clear. Corporations and government ministries are laying the groundwork for long-term, cooperative research. The scramble for market share has begun for short-term products, such as magnetic sensors, infrared sensors, linear motorcars, and thin-film process equipment. Japanese companies have shown that they can rapidly master and improve upon an existing technology, but the lack of a theoretical basis for superconductivity will challenge Japan's best scientists. Whether they can meet that creative challenge remains to be seen.

Meanwhile, the superconductor age is upon us—we are fast approaching the moment of truth. Will the United States be a major player or merely an observer in this next-generation industry? In June 1988 OTA reported to Congress:

> By and large . . . American companies have taken a wait-and-see attitude. They plan to take advantage of developments as they emerge from the laboratory—someone else's laboratory—or buy into emerging markets when the time is right. Unfortunately, reactive strategies such as these have seldom worked in industries like electronics over the past 10 to 15 years.

Japanese corporations have proven to be so quick at commercializing new technologies that a U.S. come-from-behind strategy would surely be doomed to failure. As we have seen in the television and semiconductor industries, U.S. companies that fall behind are more likely to retreat from the market altogether because of pressure from Wall Street for quarterly profits. Remaining competitive in the superconductor industry will require exactly the opposite approach. Unless the United States develops imaginative new ways to organize and finance long-term research programs, it may end up conceding superconductor industries to Japan and Europe in the twenty-first century.

13

Satori in the Laboratory: The Challenges Facing Japanese Researchers

> What should Japanese researchers do if they want to develop into first-class scientists? Right now, the only option is to go overseas. If they're really interested in science, then that's what they should do. In Japan, there are no opportunities for young researchers to do independent research.
>
> —Susumu Tonegawa
> MIT professor and
> first Japanese Nobel laureate in
> medicine and physiology (1987)

IN the early 1960s Itaru Watanabe, a biology professor working at Kyoto University's virus research institute, was approached by a promising young student who sought his advice about trying to enter the institute. Professor Watanabe, now a professor emeritus at Keio University, answered without hesitation: "It will be very difficult to pursue your research in Japan. Perhaps you should go abroad."

The young man, Susumu Tonegawa, followed his advice. He enrolled in a doctoral program at the University of California at San Diego in 1963, transferred to the Salk Research Institute where he completed his postdoctoral work in virology, then moved to the Basel Immunology Research Center in Switzerland. In 1981 Tonegawa became a professor at the Massachusetts Institute of Technology (MIT), where he continued his research on how the immune system occasionally misfunctions by producing faulty antibodies that cause allergies. In 1987 his hard work paid off. For his ground-breaking research on antibodies, Tonegawa became the first Japanese to win the Nobel Prize in medicine. In making its selection, the Nobel committee noted that Tonegawa's research was monumental because it opened the door to new methods of fighting diseases such as acquired immune deficiency syndrome (AIDS).

Tonegawa's victory was received with decidedly mixed feelings in Tokyo. Like the handful of Japanese Nobel winners who had preceded him, Tonegawa was applauded because he demonstrated to a skeptical world that Japanese are capable of making major contributions to science. The fact that he chooses to work in the West, however, is an unpleasant reminder of Japan's continuing failure to produce world-class scientists. Despite their technological strength, Japanese scientists still trail behind their Western peers. They are stifled by a rigid hierarchy, archaic rules, inadequate funding, and enormous pressures to conform. Many of the best researchers leave the country. Tonegawa's advice to young researchers touched a raw nerve, setting off yet another round of self-criticism and soul-searching about Japan's dearth of scientific creativity. In many quarters, he has become a cause célèbre for his independent thinking, but his frank criticism of Japan's hidebound scientific establishment has made him a black sheep in the family.

Speaking before a Tokyo audience, for instance, Tonegawa upset many Japanese by declaring: "Japan's science is definitely inferior to America's in terms of real creativity. It's very clear that Japan is making money by taking and applying the fruits of science that the West creates at great expense." When asked why Japanese scientists have not been more creative, Tonegawa criticizes the profit-oriented nature of Japanese government research programs. He believes the Science and Technology Agency (STA)—established in 1956 to promote national science and technology policies—does not pursue fundamental research, only applied research. This overcommercialization of scientific research has forced some of Japan's best scientists to move overseas.

While Japanese officials would like to ignore Tonegawa, he is not alone in his views. Other Japanese Nobel Prize winners are equally critical of Japan's stifling scientific environment. In fact, four of Japan's five Nobel laureates are working in the West because they find it virtually impossible to do creative research in Japan. The most notable is Leo Esaki—1973 winner of the Nobel Prize in physics for the tunnel diode—who joined IBM's Thomas J. Watson Research Center in 1960. In an interview with *Look Japan,* Esaki observed:

> You may feel much more at ease in Tokyo than New York City. But the price that you pay in such a controlled environment is a lack of creativity. Since scientific creativity is the most important element in basic research, it comes as no surprise that Japan lags behind. This is not because applied research has received undue emphasis—it is simply that basic research cannot be done . . . The Japanese do not like to hear me say this, but money is not enough: if you are not your own master, if you do not have your own soul, you will be an object rather than a creative subject.

It is this kind of stinging criticism from Japanese Nobelists that has prompted Japanese organizations to seek greater scientific creativity at home. In its 1988 White Paper, STA belatedly acknowledged: "We must pursue creative research and development, especially basic research, if we want to build harmonious international ties." Wealth is no longer enough; the Japanese are seeking international respect and, more important, self-respect. They are seeking *satori*—spiritual enlightenment—in the laboratory.

Obstacles to Scientific Creativity

Japan's scientific world is in a state of flux, especially since Tonegawa's scathing comments in 1987. Industry leaders realize they must cultivate more creative scientists and researchers, not just creative product designers, if Japan is to gain full acceptance in the world. In a 1988 survey by the *Japan Economic Journal*, 80 percent of the technology executives at 100 Japanese manufacturing companies felt Japan must strengthen its basic research facilities. The journal noted: "Japan can no longer dominate global markets by selling only improved technology and products. A sense of crisis has spread, and companies detect the need to wipe out the impression that the Japanese are enjoying a free ride on principles discovered in Europe and the United States."

However, Japan faces numerous obstacles in its pursuit of scientific creativity. Japanese researchers are still reluctant to explore the unknown alone and to challenge others at international conferences, especially Westerners. Avoiding the limelight, they prefer to follow the leaders into new fields. This cautiousness is deeply ingrained. For decades, perfecting the wheel, not inventing it, has been seen as the surest way to professional security and economic prosperity. Japanese scientists view themselves as cultivators, not as pioneers exploring new frontiers. Japan's scientific community has reinforced this conservatism by belittling the work of unorthodox researchers who stray from the mainstream. While government leaders advocate greater creativity, universities and ministries ostracize highly outspoken, independent-minded people who challenge the accepted wisdom, like Professor Tonegawa. The protruding nail is still hammered down.

Indeed, Japan's scientific establishment is a major obstacle to creativity. Professor Junichi Nishizawa of Tohoku University, inventor of the static induction transistor and other key technologies, was ignored for many years because of his criticism of Japan's basic research policies. He believes that, rather than duplicate research being done elsewhere, Japan should pursue independent research. "Research that merely monitors others' work is worthless," he says.

Until recently, his comments fell on deaf ears. Now both Japanese and Westerners agree that Japan is not contributing its fair share to the pool of basic scientific knowledge. In the past, the Japanese could plead that poverty was the reason for its weak scientific infrastructure and shortage of trained researchers, but that excuse now rings hollow. Thousands of Japanese researchers have studied and worked abroad, but are returning home to an uninspiring working environment. The Japanese economy is the second most powerful in the free world, and its capital markets and corporate coffers are bulging with enough cash to buy out U.S. high-tech companies and donate millions to U.S. and European universities for specialized research. Yet the Japanese government and university laboratories go begging for funds. It is a dismal prospect indeed for Japan's best and brightest—who usually head straight for private industry, where they have an opportunity to develop new products on state-of-the-art equipment.

There are many theories to explain why Japanese research institutions are so weak in basic research. Some critics point to the social environment, while others blame inadequate funding and the structural problems of the educational system. In his book *Japanese Technology*, Masanori Moritani, senior researcher at Nomura Research Institute, identifies three major obstacles blocking creative research in Japanese laboratories.

The biggest problem, he argues, is the college "examination hell," which rewards students adept at rote memorization and taking tests, but penalizes students with inquisitive minds. The second obstacle is Japan's single-minded "herd," or "festival," mentality (referring to the *omikoshi* parade floats carried by energetic young men), in which companies and government establish joint research projects to focus on narrowly defined goals. He believes that true creativity occurs "before the festival begins" and mourns the declining number of young Japanese researchers interested in basic research. He is pessimistic about Japan's prospects, fearing that its top researchers will be found not among the honor students but among its educational dropouts. The third obstacle facing Japan is its lack of "leeway," a "spirit of play," and a "gambler's spirit"; researchers rarely risk the possibility of failure. Moritani believes Japan's traditional R&D system will change only if there is a pressing demand for scientific creativity.

Where will this demand come from? Besides the pressure of mounting foreign criticism, Japanese laboratories are being pressured by its frustrated young people who chafe under the direction of uninspiring managers. These *shinjinrui*—the "new breed"—have swollen the work force and are reaching the point in their careers where they are beginning to press for changes in management policies.

However, Stephen Kreider Yoder, former Tokyo correspondent for the *Wall Street Journal,* observes that these young people are running into corporate walls.

> Many [of Japan's best and brightest young scientists] continue to find their creative urges stifled by a social fabric that seems to idolize seniority, loathe individualism and muffle debate. They say it is as if Japan simply won't let its youth do freewheeling science . . . The big culprits are deeply ingrained traditions, the same ones that made Japan an industrial powerhouse: acquiescence to authority, strict seniority, a stable but immobile work force and little debate. Those customs often inspire loyalty among factory hands and teamwork among the engineers who whisk product ideas to market. But they also can stunt free-ranging thought in scientists soon after they leave the university.

In his interview with Yoder, Yoshifumi Nishimura, a thirty-nine-year-old chemist, is blunt: "There's no environment where you can do wacky things that end up being creative. . . . If you're different, you're a minus. . . . That nips originality in the bud."

While social obstacles are a factor, less visible structural factors, such as inadequate funding and training, also play a major role in repressing creativity. In a 1988 *Japan Industrial Journal* survey of 1,000 businessmen, respondents ranked the following reasons for Japan's poor showing in basic research: inadequate national funding (53 percent), educational system (45 percent), corporate environment (44 percent), low status of researchers (41 percent), hyper-competition (33 percent), weak industry–university ties (29 percent), weak university research

(22 percent), lack of international exchanges (16 percent), and lack of technical information (5 percent). While the corporate environment ranks as one of the major causes, it masks other obstacles that could be overcome through changes in management and educational policies.

In this chapter, we address the major impediments facing Japanese universities, government agencies, and corporations and some of the policy measures that are currently being taken to address these problems.

University Research—The Weak Link

In the West, research universities such as MIT, Stanford, UCLA, and Cambridge are seedbeds of scientific creativity that attract the most brilliant minds from around the world. They are organized in ways that encourage frontier research. By contrast, Japanese universities are creative backwaters in the scientific world. They are severely underfunded. Research is controlled by rigid hierarchies of elite professors and government bureaucrats. Equipment is often poor and obsolete. There are few postdoctoral programs and research positions for aspiring young scientists. The lucky few who find spots in academia must wait until their late thirties before they can pursue their own research. Without these talented people, however, Japanese universities cannot attract the industry funding they need to achieve major breakthroughs. Thus, universities face a vicious cycle: lack of funding, poor equipment, a poor image, and weak industry support.

No wonder Professor Tonegawa advises young people to leave Japan if they want to pursue basic research. Their chances for success in Japan are almost nil. How are Japanese universities trying to break out of this research straightjacket as they enter the 1990s? Let's look at the major challenges facing them.

Rigid Hierarchy

A major obstacle blocking scientific creativity in Japanese universities is the *koza* (chair) system, in which professors and lab managers in their forties and fifties maintain absolute control over research topics and budgets. Numbering about 800, these *koza* managers are the pillars of Japan's scientific establishment. They advise the Ministry of Education (MOE) and STA about the direction of scientific policies and the awarding of research grants. They influence the hiring of young graduates into university and government research positions. They direct teams of young scientists, whose work appears under the names of the *kozas* and whose only hope for doing independent research is to wait patiently until they are promoted to higher posts in their late thirties. This lock-step approach to research is a key barrier to creative research. It discourages risk-taking and encourages younger people to defer to their *koza.* By contrast, talented young U.S. researchers have the opportunity to advance more quickly and can distinguish themselves as early as graduate school.

Given the dense network of social obligations among Japanese scientists, change to the system comes slowly. Young researchers may strike out on their own, as Professor Tonegawa did, but by leaving the system they find it impossible to return, thereby lessening the pressure for fundamental reform.

Inadequate Funding

Advanced research suffers at many Japanese universities because of their poor facilities and equipment. A recent survey by the Chemical Society of Japan, for example, found that half of all chemistry researchers have serious problems buying equipment that costs over 10 million yen ($83,000). Kenichi Fukui, president of Kyoto University of Industrial Arts and Textile Fibers and winner of the Nobel chemistry prize, notes that U.S. and European universities are able to conduct more creative research because they are allotted twice the research budget of Japanese universities. Until this huge disparity is corrected, Japanese universities will continue to underperform.

MOE recently established several programs to fund more basic research at national universities. In 1985 the Special Research Fellows program was created to promote independent research; as of 1987, 568 researchers had been chosen for the program. In 1987 MOE established the Key Field Research program to promote research into major technical and social issues for periods of three to six years. MOE is trying to rectify Japan's university research system with more funding, but budget ceilings make it difficult to expand research funds significantly.

Weak Postdoctoral Programs

Inadequate funding has created the problem of "overdoctoring"—a growing class of highly qualified Ph.D.s with scientific or engineering degrees who cannot find full-time work. The problem is so severe that less than 10 percent of Japan's Ph.D. candidates in the biological sciences are employed in either industry or academia upon graduation. In an unpublished paper written for Stanford University, Chris Lee describes the depressing plight of Japan's young Ph.D.s:

> Overdoctors live a marginal existence wedged indeterminately between the high status world of research science and inglorious vagabondage, waiting anxiously while the slow wheels of scientific patronage turn, in search of jobs. . . . Only through his connection to a professor (invariably, the man who advised him on his Ph.D.) can the overdoctor hope to find a position in academia or, for that matter, in industry, which is extremely reluctant to take on "new men" of such an advanced age as a Ph.D. (usually 27 or 28). It is for this reason that most overdoctors continue to work without pay in their professor's laboratory, supporting themselves through part-time jobs in cram schools while the professor uses his influence to seek out an open position.

Even at prestigious national universities such as Tokyo and Kyoto, young scientists are forced to scramble for work because of this inflexible system of academic patronage and inadequate funding.

Young researchers face other obstacles. In graduate school it is not easy to switch fields or universities. Over 90 percent of Tokyo University's Ph.D. candidates in biological sciences come from the university's undergraduate department. The twenty graduate students in each laboratory compete to become one of the four to five part-time lecturers seeking one of the three positions as associate professors, from whom only one may eventually become a full professor. It is a rigid hierarchy that gives equal weight to interpersonal skills and political maneuvering as to academic brilliance.

The result of this overdoctor situation is predictable. Listen to two top-flight Japanese researchers interviewed by Chris Lee. Dr. Hitoshi Kakidani, a graduate of Tokyo University in microbiology who is now working at a biotechnology company, says: "My brother is working for a Ph.D. in the laboratory of a professor at Kyoto University, but that professor is doing nothing to help him find a job. . . . My brother is very scared. He will become an overdoctor."

Keiichi Homma, a rebel from Japan's scientific world, is more cynical: "Overdoctor? He has very low status . . . he has no status. . . . So sometimes after two years he cannot enter a library, because he has no ID. . . . He has only his Ph.D., no position, no prospects, no job, no money, no wife, no kids . . ."

This situation is tragic not only for Japanese researchers, but for the entire scientific world. If Japan wishes to tap its latent creativity and reduce its dependency on foreign technologies, these are the people for whom a little funding would go a long way. Potential sources of funding might be Japanese corporations diversifying into new fields, technopolises opening joint research centers, foreign corporations seeking researchers for their Japanese operations, and foreign universities opening branches in Japan. It is ironic indeed that a country as wealthy as Japan wastes so much of its meager scientific talent through lack of adequate funding and support.

Weak University–Industry Ties

Corporate donations are an obvious solution to the funding crunch. Japanese companies have been generous donors to foreign universities, often funding full professorships to buy goodwill and access to top researchers. But they are still penurious with Japanese universities, which are viewed as ivory towers of limited value in pursuing new technologies.

This gap between industry and academia is rooted in competing interests. The biggest obstacle may be attitudinal; moneymaking is viewed with a great deal of disdain in academic circles. It is considered beneath one's professorial dignity to work with people more interested in profits than theoretical research. In Japan, the notion of the humble educator still captures the public imagination. Consequently, few Japanese professors work as consultants or serve on corporate boards, even those who work at private universities, which generally do not forbid second jobs.

At the national university level, MOE has sanctioned this arm's length distance between academia and industry. Close university–industry ties concern many professors who worry that their academic freedom—their right to choose research topics and freely publish the results—would be unduly inhibited by corporations more interested in maintaining industrial secrecy. This philosophy is reflected in Japanese tax laws. Corporate donations to Japanese universities are not tax-deductible as in the United States. Thus, Japanese companies usually donate equipment, but they have funded few university chairs.

Nevertheless, cooperation between Japanese universities and companies is on the rise. The Japanese government is encouraging more corporate giving. Michio Okamoto, former president of Kyoto University and chairman of the National Council on Education Reform, believes Japan needs a "closer partnership among industry, government and academe."

Since 1973, Tsukuba University professors and graduate students have worked with corporate researchers on a variety of projects. Tsukuba University and Hitachi recently developed the world's largest indirectly cooled superconductive solenoid.

In 1983 MOE established the System of Joint Research between universities and companies. Since 1984 twenty-six technopolis regions have been building research centers to promote joint university-industry-government research. In 1986 Kobe, Kumamoto, and Toyama universities were the first to open joint research centers, involving 396 projects in materials, mechatronics, and biotechnology to promote closer industry-university ties. Yet university–industry ties still remain weak or nonexistent in many regions despite local government support. In part to overcome this problem of underinvestment, the Ministry of International Trade and Industry (MITI) conceived the Regional Research Core City program in 1986 to decentralize research to twenty-eight regional cities (see Chapter 7).

In 1987, after the ceramic superconductor group led by Professor Shoji Tanaka of Tokyo University confirmed IBM Zurich's Nobel-winning superconductivity discovery, Japanese companies quickly assigned ten more researchers to the group. Companies have donated money to begin courses in superconductivity at Tokyo University and the Tokyo Institute of Technology.

In 1988 university–industry cooperation received a major boost. NEC, Nippon Steel, and NTT Corporation endowed the first professorship sponsored by industry—a chair at Tokyo University's Research Center of Advanced Science and Technology. Although still the exception to the rule, such a move signals greater corporate interest in Japanese universities. MOE also announced plans to establish fourteen industry–university joint research centers nationwide, including the five listed in Table 13–1. These centers will be completed by 1991 and will be open to foreign researchers.

Few Foreign Researchers

Western universities contain a wealth of new ideas because they are relatively unrestricted institutions, open to a constant flow of outside professors, students, researchers, and visitors. Their scientific creativity often results from the clash of opposing ideas. By contrast, Japan's government-run universities are rel-

TABLE 13-1
Joint University-Industry Research Centers Sponsored by MOE

University	Research Activities
Ibaraki	Machinery, new materials, energy systems
Utsunomiya	Mechatronics, electronic devices, artificial intelligence, human environment
Nagoya	Next-generation clean energy systems, functional crystal materials
Kyushu Industrial	Advanced manufacturing technology, measuring, new materials
Saga	Disaster and environmental engineering, precision manufacturing, biotechnology

SOURCE: Ministry of Education.

atively closed institutions, and the creative ideas that naturally permeate Western universities are noticeably absent in most Japanese universities. Foreign researchers are a rare breed. Only recently have foreigners been allowed to become full professors, although they are almost never found in administrative or policymaking positions. In most national universities, foreign students are a tiny, often neglected minority. If they come from Africa, Latin America, or Asia, they are often not welcomed by school administrators and local neighborhoods. (Japanese universities are not unique in this respect; Third World students are not welcome in many Asian universities.) As a result, many foreign students leave Japan with bitter memories of their experiences.

Language is an obstacle for foreigners, but it is not an insurmountable barrier. With adequate training and support, foreigners can become sufficiently fluent in Japanese to conduct sophisticated discussions and research in highly directed fields. Most fields have a limited vocabulary of several thousand technical terms that can be mastered within a relatively short period of time. But there are few scientific language training programs available to foreign researchers. Moreover, in Tokyo and other major cities, living costs are prohibitive because of the appreciating yen. The Japanese government is offering more housing subsidies, but this program cannot keep up with inflation and student demand. For students from poorer nations, it is becoming financially impossible to study in Japan.

This situation, however, is also changing. Foreign universities are trying to break into the hermetic Japanese educational system. Recently, U.S. universities have approached MOE and local governments for permission to open Japanese branch campuses. The best known is the new Kyoto Center for Japanese Studies (KCJS), a nine-month academic program run by Stanford on behalf of a nine-member university consortium that includes Brown, Chicago, Columbia, Cornell, Harvard, Michigan, Princeton, Stanford, and Yale. Table 13–2 lists some of the other U.S. universities that have established Japanese campuses; more campuses are being negotiated. The negotiations are time-consuming because of MOE's concern over differing standards and the impact of the foreign presence on Japanese universities. As Japan globalizes, however, the pressure to open its educational system to foreign scholars will increase, not decrease. If Japan seeks to achieve global stature in university research, it has no choice but to open its closed system of education.

Even if the Japanese government invites more foreigners to Japanese universities, some critics believe that more funda-

TABLE 13–2
New Foreign University Campuses

University	Location	Opening
Temple University	Tokyo	1982
American Universities League	Yokohama	1987
University of Nevada, Reno	Tokyo	1987
Southern Illinois University	Nakajo, Niigata	1988
Ohio University	Komaki, Aichi	1990[a]
Texas A&M University	Koriyama, Fukushima	1990[a]
Texas International Educational Consortium	Kashima, Ibaraki	1990[a]
Mississippi State University	Omachi, Nagano	1991[a]

[a]Tentative contracts
SOURCE: Izumi Oshima, "New College Towns Cheer U.S. Universities," *Japan Economic Journal* (8 October 1988): 32.

mental changes are required. In an interview with Malcolm Debevoise, Nobel laureate Leo Esaki of IBM says:

> Bringing in foreign scientists, encouraging collaboration between universities and private industry—these things are well and good, but they do not amount to a solution. You cannot separate scientists and research from society. In reforming education, in particular, you have to come to terms with the Japanese character. . . . In Japan, discipline means that you have to defer to senior people; you have to bow; you have to use polite words; you have to agree—a perfect example of heteronomy, of being directed by others. The problem is how to change the cultural habits of a nation.

Government Research—A Closed System

At the apex of Japan's research community are the national government researchers who, like their university counterparts, live in a closed, highly stratified society that pays more deference to seniority and rank than to original research. Usually recruited from Tokyo, Kyoto, Osaka, and other top national universities, these elite researchers form the core of Japan's next-generation research programs. The majority of them work in the Tsukuba Science City, which is located thirty-five miles northeast of Tokyo in rural Ibaraki Prefecture. In ninety-one national and private laboratories in this "city of brains," the Japanese government has concentrated over 12,000 researchers who are busy exploring frontier technologies ranging from fine ceramics, bioengineering, and advanced robotics to supercomputing, earthquake prediction, pollution control, and optomechatronics. At first glance, Tsukuba's brainpower is impressive, comparable perhaps to what a similar concentration of America's top government researchers into a single science city near Washington, D.C., would be. But here the parallel ends.

Isolated Researchers

Researchers at Tsukuba face serious obstacles in their efforts to produce creative research. Perhaps the biggest problem is their isolation. Cloistered in their separate laboratories, these researchers lead a highly compartmentalized existence, with nothing like the rich cross-fertilization of ideas that occurs more freely in most U.S. and European national laboratories.

Japanese researchers are also far removed from the centers of Western scientific excellence. They rarely sponsor foreign researchers in their laboratories and do not mingle freely with colleagues from other ministries. In Tsukuba, for example, there are few common meeting halls, and most research is highly structured along ministry lines. Multidisciplinary research is usually limited to formal exchanges between ministries. Moreover, because lifetime employment is common and there is rivalry among ministries for limited funding, researchers usually keep to their own groups. To escape this isolation, Tsukuba researchers must go to Tokyo to participate in the rich array of technical conferences and meetings that can be found there.

Recently, Japanese ministries have established several programs to stimulate more interaction among researchers. The Agency for Industrial Science and Technology (AIST), which runs MITI's laboratories, sponsors annual conferences in Tsukuba to update researchers on the status of their programs. In addition, four researcher exchange programs have been established. MITI runs the Foreign Researcher Invitation program. STA sponsors 100 foreign researchers in Japan for six months to two years through its International Basic Research program. More than thirty foreign researchers have worked at STA's program, Exploratory Research of Advanced Technology (ERATO), over the years. And MOE offers 100 postdoctoral fellowships to foreigners through the Japan Society for the Promotion of Science (JSPS).

Since 1986, the Japanese government has entered into agreements with leading industrialized nations to promote international cooperation in eighteen research projects ranging from biotechnology, optical compounds, and solar energy to new materials, remote sensing, and nuclear fusion. The Human Frontier Science Project, a long-term program run jointly by MITI and STA, is seeking international cooperation to pursue research on the human brain and the biological sciences. MITI's International Superconductivity Technology Center (ISTEC) has attracted five foreign companies. The Council for Science and Technology has formed an international research program devoted to new materials, such as fine ceramics, high-purity crystals, and biochips.

So far, there has been limited foreign interest in these programs because of their high membership fees and "first-to-patent" rules in Japan (versus "first-to-invent" in the United States and Europe). Moreover, foreign researchers are reluctant to live in Japan because of the problems with language and cultural barriers, housing, living costs, family adjustment, and education for their children; they also worry about falling behind their Western peers. In 1989 STA sought 100 foreign researchers, but ended up accepting only seventeen; more than 300 Japanese researchers work at the National Institutes of Health in Bethesda, Maryland. The number of foreign researchers in Japan is gradually increasing, but they are still a rare breed.

Cutting Through Government Red Tape

Like governments everywhere, the Japanese government imposes strict rules on researchers that dictate everything from equipment purchases to research trip authorizations. Japanese researchers, however, labor under tighter restrictions than most. Grant proposals, for example, are often delayed or ignored if they fall outside the mainstream of research or cannot promise immediately useful applications. To stretch limited budgets, laboratory officials are forbidden to dispose of obsolete equipment and replace it with state-of-the-art equipment. Even until 1984, corporations were prevented from donating equipment to MITI laboratories because of concern over undue influence in the setting of research priorities. (Heated competition among Japanese companies often threatens to sidetrack basic researchers into more commercializable research.) While this policy has protected the integrity of basic research, it has weakened government-industry cooperation and reduced the opportunities for creative research.

This situation is gradually changing. Since MITI's call for more creativity in its famous "Visions for the 1980s," Japan has dramatically increased its spending on advanced research. Total public and private R&D spending leaped from 1.95 percent of the GNP in 1975 to 2.77 percent in 1985, surpassing the United States, which increased its overall R&D spending from 2.2 percent to 2.72 percent during the same period. In 1986 total Japanese R&D spending was $70.8 billion, compared with $148 billion for the United States. The Japanese government plans to increase total R&D spending to 2.8 percent of the GNP by 1990, and to 3.0 percent by the year 2000.

Despite the increased R&D spending, the government's role in advanced research has been declining in recent years because of budget constraints. In 1985 Japan spent only 2.9 percent of its national budget on science and technology research, while the United States and West Germany spent 5 percent, and France 6.9 percent. Since budget ceilings were set in 1982, Japanese government research projects have been poorly funded. Most projects are unable to fund fully equipped research laboratories; they can only maintain administrative offices in Tokyo and must rely on the private sector to meet most of their equipment and actual research costs. In 1985 the Japanese government accounted for only 19.4 percent of total R&D spending in Japan, compared with government contributions of 46.8 percent in the United States and 39.8 percent in West Germany. It should be noted, however, that Japanese government R&D spending is heavily focused on civilian research because of Japan's small defense research budget.

Basic research in Japan is primarily conducted under the auspices of twenty-three government ministries and councils. In fiscal 1987 (1 April 1987 to 31 March 1988), Japan's total science and technology budget was about $13.8 billion. The largest budget allocations went to the MOE ($6.5 billion), STA ($3.5 billion), and MITI ($1.8 billion).

STA is responsible for maintaining the Tsukuba Science

City. Since the early 1980s, STA has established several programs to promote creativity in frontier research.

For example, the Japan Research and Development Corporation (JRDC), which was founded in 1961 and as a statutory corporation under STA, was one of the early initiators of Japan's push into basic research. In 1981 JRDC established the ERATO program to promote creative scientific research among young scientists and engineers. Table 13–3 lists fifteen ERATO projects that are underway or have been completed. Nine of these projects have been started since 1984.

These programs are usually directed by university professors or corporate research managers who are interested in pursuing unexplored fields of biological and physical science. Unlike in past Japanese programs, the project directors are free to choose challenging, open-ended research themes and to recruit their own researchers, usually young Ph.D.s in their early thirties from various government, university, and corporate labs around the world.

The Nishizawa Perfect Crystal Project (1981–86) is an example of JRDC's impact on critical emerging technologies. Directed by Professor Junichi Nishizawa of Tohoku University, the project focused on developing new semiconductors by creating defect-free gallium arsenide (GaAs) crystals, thereby overcoming a major obstacle in the production of high-quality GaAs chips. These compound semiconductors can be used for ultrafast supercomputer logic circuits, high-speed switching, and optical functions. By the project's end in 1986, Dr. Nishizawa's team had developed an arsenic pressure-controlled method for developing GaAs single crystals using software programs to automate the crystal growth process. With this key technology, Japanese companies such as Sumitomo Metals and Dowa Mining were able to develop GaAs crystals for use by Fujitsu, Hitachi, and NEC—Japan's top supercomputer makers developing GaAs chips for their next-generation machines.

Since October 1986, STA has also pursued the Frontier Research program through its Institute of Physical and Chemical Research (RIKEN). Projects in this program generally run for fifteen years, are open to foreign scientists, and are in one

TABLE 13-3
JRDC's ERATO Program

Period	Research Theme	Research Director
1981-86	Ultrafine Particle	Dr. Chikara Hayashi President, ULVAC
1981-86	Amorphous and intercalation compounds	Dr. Tsuyoshi Masumoto Professor, Tohoku University
1981-86	Fine polymers	Dr. Naoya Ogata Professor, Sophia University
1981-86	Perfect gallium arsenide crystals	Dr. Junichi Nishizawa Professor, Tohoku University
1982-87	Bioholonics	Dr. Denichi Mizuno Professor, Teikyo University
1983-88	Bioinformation transfer	Dr. Osamu Hayaishi President, Osaka Medical College
1984-89	Superbugs	Dr. Koki Horikoshi Chief Scientist, RIKEN
1985-90	Nano-Mechanisms	Mr. Shoichiro Yoshida Managing Director, Nikon
1985-90	Solid surfaces	Dr. Haruo Kuroda Professor, Tokyo University
1986-91	Quantum magneto flux logic	Dr. Eiichi Goto Professor, Tokyo University
1986-91	Molecular dynamics assembly	Dr. Hirokazu Hotani Associate Professor, Kyoto University
1986-91	Biophoton	Dr. Fumio Inaba Professor, Tohoku University
1987-92	Terahertz	Dr. Junichi Nishizawa Professor, Tohoku University
1987-92	Morphozenes	Dr. Mitsuru Furusawa Deputy General Manager, Dai-ichi Seiyaku Co., Ltd.
1987-92	Molecular architecture	Dr. Toyoki Kunitake Professor, Kyushu University

SOURCE: Japan Research Development Corporation.

of two fields: bio-homeostatis (aging processes, intestinal organisms, and plant biology) and frontier materials (bioelectronic, quantum, and nonlinear optical materials). Four laboratories have been created to pursue research on bio-homeostatis: the

Molecular Regulation of Aging Laboratory (genetic research), the Aging Process Laboratory, the Intestinal Flora Laboratory (the relationship between aging and stomach bacteria), and the Plant Biological Regulation Laboratory (molecular and cellular mechanisms). For research on frontier materials, RIKEN set up the Quantum Materials Laboratory, the Nonlinear Optics and Advanced Materials Laboratory, and the Bioelectronic Materials Laboratory. In 1987 RIKEN spent $110 million on these laboratories.

RIKEN has stimulated private investments in genetic and protein engineering. For example, fourteen companies led by Mitsubishi Chemical Industries and Kyowa Hakko Kogyo formed the Protein Engineering Research Institute (PERI) at Senri New Town in Osaka in 1988. The institute is conducting basic research on the artificial synthesis of proteins for use in future pharmaceuticals and biocomputers. Because of its strategic nature, foreign companies such as Nihon Digital Equipment Corporation and Nippon Roche K.K. have joined as charter members.

STA has also set up the National Information System for Science and Technology (NIST), which is developing a Japanese–English computer translation system to speed the flow of technical information to corporations. The program is coordinated by the Japan Information Center of Science and Technology (JICST), which has operated an on-line information system (JOIS) since 1976.

Through its Agency for Industrial Science and Technology (AIST), MITI is sponsoring large-scale R&D projects in fields such as fifth-generation computing, high-performance ceramics, and advanced robotics (see Table 13–4). These projects focus on basic technologies and generally run ten years, much longer than previous Japanese government projects. Kazuhiro Fuchi, head of the Institute of Next-Generation Computer Technology (ICOT), notes that it generally takes ten years for basic research on a given project, and five more to create a commercial product. Many foreign observers have already written off ICOT's Fifth Generation Computer Project because of the lack of short-

TABLE 13–4
Japan's Major Large-scale R&D Projects

Project	Organization	Period	R&D Budget ($ millions)
Fifth-generation computer	Institute of Next-Generation Computer Technology (ICOT)	1982–1991	$416
High-performance ceramics	Engineering Research Association for High Performance Ceramics	1981–1992	50
Advanced robot technology	Advanced Robot Technology Research Association	1983–1990	167
Three-dimensional integrated circuits	Research and Development Association for Future Electron Devices	1981–1990	41[a]
Telecommunications basic technology	Advanced Telecommunications Research (ATR) Institute International	1986–1995	583
Protein engineering	Protein Engineering Research Institute	1986–1995	142
Ultra pioneering processing system	Advanced Material Processing and Machining Technology Research Association	1986–1993	125

[a]Up to fiscal 1987
NOTE: $1 = 120 yen
SOURCE: Ministry of International Trade and Industry.

term results. What they overlook is that Japanese companies are already using AI and expert systems in medical systems, factory automation, chip design, and financial credit analysis. Major commercial spin-offs will not occur until the early 1990s.

One new large-scale project at the Osaka-based ATR Interpreting Telephony Research Laboratories is extremely ambitious. An automatic translation telephone is being developed on which users can speak to each other in different languages. Recently, ATR invited researchers from Carnegie-Mellon University and several French universities to join the project. The British government is very interested and has agreed to sponsor joint research with Japan in the field of telephone translation systems. With the globalization of business, many countries realize that the nation that dominates automated telephone translation technology will reap enormous revenues and profits in the future.

In 1986 MITI drew up a scenario for high-tech industries as a stimulus to private investment. According to its projections, the Japanese electronics market will reach $265.8 billion, new materials $45 billion, and biotechnology $41.7 billion by the year 2000. Related markets would add another $1.9 trillion and 1.17 million workers. While these markets may eventually materialize, MITI's overly optimistic forecast is typical of Japanese government reports. Many Japanese executives observe that research in biotechnology and fine ceramics has yet to yield much profit. Even MITI officials acknowledge their boosterism. "Dreaming is over," says Naomichi Suzuki, director general of MITI's Basic Industries Bureau. "We are now pursuing goals in a controlled manner while we used to paint vague pictures."

Key Technology Center

Besides the large-scale research projects, the Japanese government has begun to take an entrepreneurial role in promoting creative technologies. In 1985 MITI and the Ministry of Posts and Telecommunications jointly established the Key Technology Center (KTC) (Kiban Gijutsu Kenkyu Sokushin Senta). The center's mission is to promote basic research in the key technologies needed for the leap to next-generation industries. It acts as an "investment banker" in the sense that it funds risky, leading-edge research companies or consortia. As shown in Figure 13–1, KTC is very different from previous Japanese national research projects. Whereas MITI receives its funds from the national budget, KTC is funded by the Japan Development Bank, private industry, dividends from the sales of Nippon Telephone and Telegraph Corporation (NTT) and Tobacco Corporation stock, and dues paid by government banking institutions. KTC invests these funds into research programs proposed by individual or multiple companies seeking to conduct research. In principle, KTC funds 70 percent of the project cost,

FIGURE 13–1
Key Technology Center Funding

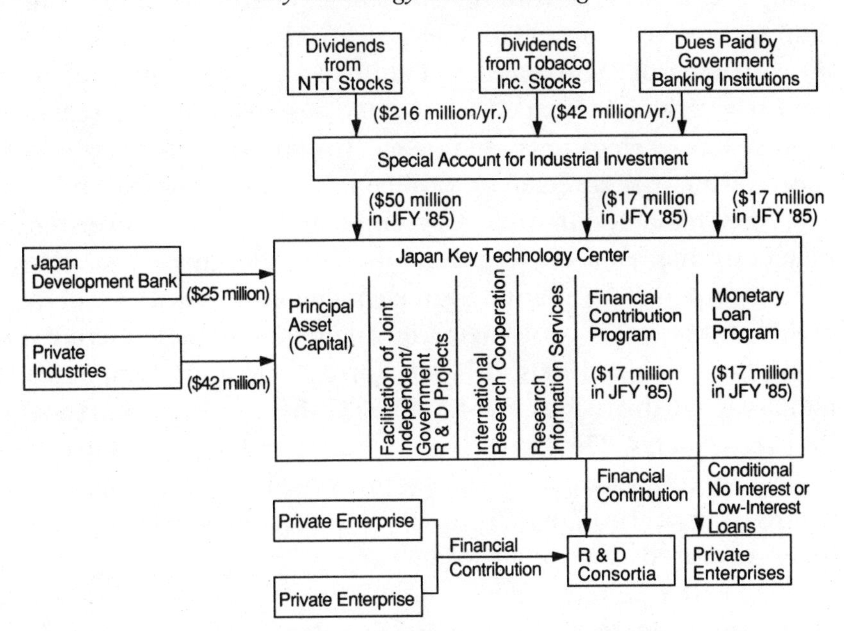

SOURCE: National Science Foundation, Tokyo Office.

with private companies funding the remaining 30 percent through a legal holding company.

Since its founding, KTC has funded over 200 projects. Some of the research topics covered include protein and genetic engineering, nonoxidized glass for optical memory systems, optocomputing, two-way television, electronic dictionaries, natural language translation phones, communication networking, sight and hearing research for better human/machine interface, and intelligent building construction technologies. These projects generally involve from five to twenty companies and up to 100 dedicated researchers.

An example of a KTC project is Sortech, a joint venture of thirteen Japanese semiconductor device and equipment makers. Formed in June 1986, Sortech is supported by the top chip makers in Japan, including Canon, Fujitsu, Hitachi, Matsushita, Mitsubishi, NEC, Nikon, Oki Electric, Sanyo, Sharp, Sony, Sumitomo Electric, and Toshiba. Their goal is to develop compact synchrotron orbital radiation (SOR) equipment for manufacturing memory chips with densities of sixty-four megabits (one million bits) or more. (The computer industry now uses one-megabit memory chips and will shift to four-megabit chips in the early 1990s.) SOR technology involves developing circular or oval-shaped rings to accelerate electron particles to very high speeds, which are directed onto silicon wafers to draw extremely thin lines for chip circuitry. Currently, SOR rings are over ten meters in diameter, making them much too large for practical use in chip-making plants. Sortech members plan to develop SOR rings less than one meter in diameter for their memory chip plants in the 1990s. In 1988 Sortech member companies had assigned thirty-four full-time researchers to their Tsukuba Research Center. Their goal is to lay the foundation for Japan's push into large-capacity memory chips, which will be used in future computers, high-definition TVs, facsimile machines, and other advanced electronic products.

Thus, the Japanese government is trying to overcome the numerous obstacles facing the growing demand for a more creative research environment. The ministries harbor no illusions about the difficulty of promoting creativity; they realize it is not something that can be mandated by law. By seeding new projects with adequate funding, however, they hope to allow private and public researchers to make significant breakthroughs.

The Boom in Corporate Research

For many Westerners the notion of creativity in a Japanese corporation is an oxymoron. They cannot imagine how workers in Japanese companies could possibly overcome the strong pressures to conform. Management experts who have studied the *ringi* method of building consensus and harmony through *nemawashi*—binding the roots—conclude that it stunts individual creativity. They note that the corporate hierarchy and dependency relationships (*amae*) prevent researchers from contradicting their seniors. Yet, these theories do not explain how Japanese companies are able to develop very creative new products, such as Funai Electric's automatic bread-baking machine and Temaki's paper-thin audio speakers.

Indeed, leading Japanese companies display an uncanny ability to innovate, despite the many obstacles facing them. How do they do this? How are they repositioning themselves to become more creative? In this section, we review six strategies being pursued by Japanese corporations: new research centers, aggressive recruiting, liberalized R&D management policies, corporate spinouts and buyouts, strategic alliances, and donations to foreign universities.

New Research Centers

Since the yen shock of 1985, Japanese companies have beefed up their research capabilities to compensate for the declining competitiveness of their products, and as a result, there has been a boom in new research laboratories and software centers. According to the *Japan Economic Journal,* about 160 private R&D centers were built from 1985 to 1987, of which 40 percent are located in the Tokyo region. More than 100 of these R&D centers have been established by Japanese and foreign companies in the eight industrial parks surrounding the Tsukuba Science City. The other R&D centers are clustered in neighboring Chiba, Kanagawa, and Saitama prefectures, and in Kobe, Kagawa, and Hiroshima.

Since 1987, Japanese R&D spending has picked up substantially. In 1988 the *Japan Economic Journal* surveyed 177 Japanese companies and discovered that most have stepped up their overseas R&D efforts, especially in the United States and Europe. Of the sixty-six foreign R&D centers opened by fifty-eight Japanese companies, twenty-three were begun in 1987 and 1988. Moreover, thirty-seven companies were planning new overseas R&D centers. Table 13–5 shows recently opened R&D centers overseas. Executives mentioned three key factors in their decisions to open R&D centers abroad: being close to their customers, getting access to foreign technology, and hiring foreign R&D talent.

Dataquest found that electronics makers accounted for over

TABLE 13–5

Overseas Japanese R&D Centers

Company	Open	Location	Activities
Aisin Seiki	1987	Sophia Antipolis, France	Electronics
Asahi Optical	1985	Englewood, Colorado	Optical disks
Canon	1988	England	n/a
Canon	1990	Newport News, Virginia	Copiers, office automation design
Hitachi	1985	Atlanta, Georgia	Telephone exchange software
Honda Motors	1984	Los Angeles, California	Electronics, automotive
Horizon Research	n/a	Waltham, Massachusetts	Computers, software, systems
Isuzu	1987	Los Angeles, California	Automotive research
Kanto Electronics	1984	Silicon Valley, California	Integrated circuits, peripherals
Kao	1986	West Berlin, West Germany	Haircare, shampoo, rinses
Kao	1986	Los Angeles, California	Haircare, shampoo, rinses
Kao	1987	Barcelona, Spain	Haircare, shampoo, rinses
Kobe Steel	1986	North Carolina	New materials, electronics
Kobe Steel	1988	London, England	Research center
Kyocera	1985	Vancouver, British Columbia	Electronic materials
Matsushita	1976	Palo Alto, California	Analog and digital custom integrated circuits
Matsushita	1981	Burlington, New Jersey	Video broadcasting systems
Matsushita	1981	Santa Barbara, California	Speech acoustics, synthesis
Matsushita	1986	Taiwan and West Germany	Electronics
Matsushita	1987	Woodside, Illinois	Point-of-sales systems, computers, industrial electronics
Mazda	1987	Detroit, Michigan	R&D center
Mazda	1988	Irvine, California	Automobile trends
Mitsubishi Electronics	1984	Durham, North Carolina	Semiconductors
Nakamichi	1984	Torrance, California	Nonaudio electronics

Nakamichi	1988	Mt. View, California	Disk drives
NEC Electronics	1986	Sunnyvale, California	Very large scale integration research
NEC Home Electronics	1987	Mt. View, California	PCs, home electronics
NEC Memory Design Center	1988	Natick, Massachusetts	Specialty memories
NEC Research Institute	1988	Princeton, New Jersey	Human thinking/recognition; knowledge representation; audiovisual information structures and processing; new devices
NEC Systems	1985	Foxboro, Massachusetts	Software, prototype hardware
Nippon Densan	1988	Kansas	n/a
Nippon Denso	1986	Battle Creek, Michigan	Automotive electronics
Nissan Motors	1983	Michigan	Automotive
Otsuka Pharmaceuticals	1982	Frankfurt, West Germany	Pharmaceuticals
Otsuka Pharmaceuticals	1985	Rockville, Maryland	Cytology, immunology
Photonic Integration Research (NTT)	1987	Columbus, Ohio	Optoelectronics
Ricoh of America	1963	San Jose, California	Copiers, fax, office automation
Ricoh Scientific Telecommunications Systems	1979	Santa Clara, California	Fax equipment
Ricoh Systems	1979	San Jose, California	Application-specific integrated circuits, complementary metal oxide silicon wafers
Shinetsu	1987	California	Technical services
Sony	1984	Portland, Oregon	Optical recording media
Sony Research Center	1988	Palo Alto, California	NEWS workstation
Sumitomo Electric	1983	Raleigh, North Carolina	Optoelectronics
Toyota Motors	1977	Los Angeles, California	Automotive
Yaesu Radio	1985	California	Satellites, radios

SOURCE: Dataquest, Inc.

100 of the R&D centers established since 1984. These research labs are focusing on leading-edge technologies such as megabit memory chips, voice and image recognition, artificial intelligence, video image processing, neural networks, and biocomputers.

Japanese companies are newcomers to the international "brain gain" market because of their relatively closed hiring practices and their late start in globalizing their operations. But the race to develop creative products is forcing them to open research centers abroad to tap foreign brainpower and escape the R&D straitjacket at home.

Hitachi, the world's fifth largest electronics company, has an ambitious worldwide R&D strategy. In 1988 it spent $2.4 billion on R&D, or almost one billion dollars more than the National Science Foundation. Hitachi's R&D organization is massive: it has 16,000 researchers working in twenty-two laboratories. At any moment, more than forty-five research projects are underway, such as neural computers, superconductivity, language translation machines, high-definition displays, and large color liquid crystal displays (LCDs) for televisions and computers. Most of the research focuses on short-term payoffs (one to five years), while the Advanced Research Lab looks out twenty to thirty years. In 1988 Hitachi expanded its overseas R&D network by opening four R&D centers: two in Europe (one at Cambridge University's Cavendish Laboratory to develop high-speed chips, and one at Dublin University's Trinity College for supercomputer software), and two in the United States (one in Silicon Valley to design high-speed signal processing chips, and one in Detroit for car electronics). Hitachi Metals opened a magnetic materials development subsidiary in California in early 1988.

Other Japanese companies are also expanding their overseas R&D networks. Fujitsu opened two R&D centers in California and one in North Carolina, plus five semiconductor design centers in the United States. Its Silicon Valley design center developed the SPARC microprocessor for Sun Microsystem's popular technical workstations after Sun was turned down by U.S. chip makers. In 1988 Sony followed suit by opening a software development center in Silicon Valley for its NEWS technical workstation, followed by a new video technology center in 1989 for servicing HDTV cameras and studio equipment for the U.S. market. Meanwhile, Matsushita has quietly opened four R&D centers in the United States to maintain its lead over Sony in the U.S. video market.

To date, NEC Corporation has made the boldest move. In July 1988 the NEC Research Institute was established in Princeton, New Jersey, to develop an "interpretation telephone" for translating natural languages, such as Japanese into English and vice versa. To achieve this goal, NEC is conducting research in areas such as modeling human thinking and recognition, knowledge representation, information structures, visual/audio data processing, and new device concepts. According to Dr. Michiyuki Uenohara, R&D director at NEC, the center is devoted to basic research and its activities will be open to joint research with universities and other organizations. NEC plans to spend $77 million over the first five years and to eventually employ 100 researchers.

Japanese companies are also sponsoring global think tanks to address long-term environmental and economic development issues. In November 1988, for example, Tokyo Electric Power Company's chairman, Gaishi Hiraiwa, was appointed to head a global industrial and cultural research center sponsored by Nippon Steel, Toyota Motor, Sony, Mitsubishi, the Industrial Bank of Japan, and Tokyo Electric Power. The institute, suggested by former MITI Vice Minister Shinji Fukukawa, is focusing on global environmental and resource problems such as acid rain, the "greenhouse effect," population growth, and international conflict resolution. Japan's pursuit of these fields of research not only will fulfill some of Japan's global responsibilities but will create new commercial markets, such as better smokestack scrubbers, less-polluting automobile engines, and new weather forecasting supercomputer programs.

Multinational R&D Labs in Japan

The Japanese are not the only ones opening research centers overseas; European and U.S. companies are also boosting their

R&D spending in Japan to acquire leading-edge technologies and develop products for the Japanese market. Since 1985, nearly twenty European and U.S. chemical, electronics, and pharmaceutical companies have opened R&D centers in Japan (see Table 13–6). By working closely with Japanese customers, they are tailoring their products to meet local quality expectations, government licensing requirements, and after-sales servicing demands.

IBM Japan, which accounts for about 10 percent of IBM's total sales, reacted to growing Japanese R&D spending by expanding its three R&D Japanese laboratories in 1988. Its Tokyo Research Laboratory, which focuses on voice recognition and synthesis, translation, artificial intelligence, image processing, graphics, and robotics, will grow from 1,600 to 3,000 researchers. The Yamato Development Laboratory went from 80 to 130 researchers to develop new personal computers and *kanji*-based Japanese language word processors. The Yasu

TABLE 13–6
New Foreign R&D Centers in Japan, 1986–89

Year Open	Company	Location	Research Focus
1985	Texas Instruments	Tsukuba	Semiconductors
1985	Data General	Koda	Minicomputers
1985	Intel	Tsukuba	Integrated circuit design center
1985	IBM Japan	Daiwa	Computers, office automation
1987	DuPont	Yokohama	Full-scale R&D center
1987	Imperial Chemical Industries (ICI)	Tsukuba	Engineering plastics, optoelectronics
1987	LSI Logic	Tsukuba	Semiconductors
1988	DuPont	Tsukuba	Agricultural chemicals
1988	Eastman Kodak	n/a	Photography
1988	Bayer	n/a	Pharmaceuticals
1988	IBM Japan	Yasu	Semiconductors
1988	IBM Japan	Chiba	Software center
1988	Deutsche Bank	Kanagawa	30-firm incubator
1988	Schlumberger	Kanagawa	Machinery
1988	Henkel Hakusui	Ibaraki	Chemicals
1988	Japan Univac	Tokyo	Artificial intelligence
1988	W.R. Grace	Atsugi	Ceramics, electronics
1989	Ciba-Geigy	Hyogo	Pharmaceuticals
1990	Hoecht	Saitama	Agricultural sciences

Applied Technology Laboratory was opened in May 1988 to develop semiconductors and mounting technology.

Multinational companies are also bringing their Asian customers to Japan-based research centers to tap into the local technology pool. Many Asian researchers attend Japanese technical conferences to keep current with state-of-the-art advances. Apple Computer, for example, often holds technical and marketing meetings in Tokyo for its Asian educational marketing staff.

Many multinational companies are concerned over being left behind in the 1990s. They are recruiting Japanese talent and seeking access to government labs and universities in order to monitor Japanese technologies. Japanese companies dominate many key technologies, such as ceramic packaging, fermentation, LCDs, memory chips, semiconductor lasers, and optoelectronic devices that will be used in next-generation computers and telecommunications systems. Gradually, the one-way flow of technology to Japan is becoming a two-way flow as more foreign companies open research centers in Japan.

Hiring Creative People

Hiring has always been a fiercely competitive event in Japan, centered around luring graduates from top-ranking public and private universities such as Tokyo, Kyoto, Waseda, and Keio universities. The infamous "examination hell" leading to the college entrance examination, followed by company entrance exams and interviews, enables companies to pick smart, hardworking young people who can be molded into company team players. This hiring process, however, screens out creative people with weak academic records and those who do not "fit the mold."

To overcome this conservative bias, leading Japanese companies are modifying their hiring policies. Increasingly, there is a growing demand for people with special talents: mavericks with unique ideas, Japanese returning from abroad, and Japanese graduates of foreign universities. Some companies are recruiting actively at second- and third-tier universities and technical colleges, from which most of Japan's entrepreneurs have emerged in the past.

Isuzu Motors, for example, seeks people with individuality. In their interviews with candidates, Isuzu managers keep their eyes open for different kinds of people who might give Isuzu a creative edge: nonconformists, people with one or two areas of expertise, intensely curious people, and strong-willed people with novel ideas.

Office equipment maker Alps Electric, on the other hand, specifically avoids hiring five types of people it believes lack the potential for genuine creativity: those who have never failed, those with brilliant educational backgrounds, those who always behave well, those who only want to live in Japan, and those who are not attuned to their surroundings.

Nintendo, the highly successful developer of "Super Mario Brothers" and other video games, deliberately seeks "oddballs" who lack proper credentials from top-ranking universities. Many of its best software programmers are high school graduates or college dropouts who spend a lot of time reading comics, playing video games, and pursuing other hobbies.

Soichiro Honda of Honda Motors, a member of the Buddhist-based Mukta Research Institute, is especially interested in finding and nurturing creative people. Honda Motors follows five principles in recruiting and promoting its people. First, its employees are encouraged to polish their skills and work for themselves, not for the "good" of the company. Second, they are reminded that if they have talent, it should not be hidden (Reversing a famous Japanese proverb: "The eagle hides its talons"). Third, Honda believes one should not be afraid of failure. Fourth, employees are sensitized to business realities—the realities of place, things, and circumstances. And finally, they are reminded not to copy others, but to do something unique. According to Hideo Tsugiura, director of Honda's research laboratory, Honda entrusts its younger staff with a high degree of responsibility and encourages competition among them.

In 1978 Honda assigned a team of young researchers, all less than thirty-five years old, to design a "mini" Civic that would be 100 millimeters shorter than the original Honda Civic. The team consisted of people with widely different backgrounds. Indeed, except for being "car maniacs," the team was so diverse that many Honda managers worried that it would be unable to reach consensus. Yet, by channeling the diverse energies of the team members, Honda developed a novel engine, suspension, and radial tire designed specially for the new "City" model. Moreover, the team members applied for ninety patents during the project.

How did Honda achieve this? Writing in the *Sloan Management Review,* Professor Ikujiro Nonaka of Hitotsubashi University noted that "Mr. Watanabe [the team leader] held day-and-night discussion sessions, most often in rooms at small taverns or inns near the research lab instead of at luxurious hotels. In ceaseless discussions from morning to midnight, participants had to use all their wits to challenge existing paradigms."

These companies are the wave of the future in Japan. No longer content with yes-men, they are seeking new ways to challenge and stimulate their young people. These companies are still exceptions to the rule. If "necessity is the mother of invention," however, global competition leaves other Japanese companies no choice but to follow suit.

Head-hunting Japanese-style

The handwriting is on the wall. After decades of company loyalty and lifetime careers from its employees, Japanese companies are raiding each other for talented and creative scientists, managers, and software programmers. The dizzying pace of technological change has altered the rules of the game. No longer can companies rely on internal talent; they must seek fresh ideas from outside. Within the last few years, head-hunting, once a corporate taboo, has become a fast-growing business in Japan. Japanese companies are turning to executive recruitment and outplacement firms to find the talent they need. Today, there are over eighty firms recruiting new employees from the ranks of job-hopping professionals.

Hiroyuki Koyama, senior staff writer of the *Japan Economic Journal,* observes:

> Whether it's iron and steel companies looking for expertise to help them move into electronics, breweries recruiting biotechnology experts, or Japan Railway firms seeking guidance away from the tracks, everywhere the story is the same: firms are no longer hesitating to poach the kind of mature talent they need to advance and diversify.

As always, the competition is intense. Tatsuo Totani, president of Interface Consultants, a recruiting firm, says that companies are becoming aggressive in their search. "Head-hunting by designation—naming names—is picking up steam. One in three clients who wouldn't mention the names of people they want in other companies do so now, and their number is growing."

In 1987 the Ministry of Labor surveyed 4,100 companies and discovered that 25 percent of all machinery companies and nearly 40 percent of all consumer product companies had hired experienced engineers and researchers away from other companies. Computer makers, retailers, and financial institutions have been particularly aggressive. In 1987 Seibu Department Store led the pack by hiring 235 mid-career recruits, followed by Nikko Securities with 160 and Mitsubishi Bank with 40. In 1988 IBM Japan topped them all by recruiting 500 mid-career engineers and managers.

Japanese companies are rigorous in their screening, but probably none is more demanding than Nissan Motor, which is recruiting heavily to expand into car electronics, aviation, space development, and marine businesses. Nissan hires selectively, using a three-stage selection process to find technical experts. Candidates are screened by written examinations, personal interviews, and peer review by the technical staff. During the first half of 1988, only 40 of the 500 applicants were offered jobs at Nissan. These newcomers represent the cream of the crop, Japan's new wave of job-hopping engineers.

As in the United States, head-hunting is forcing Japanese companies to protect their technology and business secrets more carefully. In 1987 Fujitsu asked its employees to sign a nondisclosure agreement in the event of their being hired by another firm. Fujitsu is particularly concerned about protecting chip wiring techniques, manufacturing know-how, and intellectual property—both its own and those of its partner, IBM Japan. As recruiting activity picks up, more Japanese companies will resort to nondisclosure agreements, patent protection laws, and intellectual property laws to protect their technologies. Already, Japanese companies are filing lawsuits against imitators, such as NEC's recent suit against Seiko Epson, which sells NEC-compatible personal computers. Ironically, the United States, which is worried about losing its technologies to Japan, is the legal model for Japanese companies worried about losing their technologies to other Japanese and Asian companies.

Even among close business partners, the loss of talented people to competitors can quickly affect corporate relationships. Takashi Takagaki, manager of AMD Japan's semicustom chip design center, relates what happened when a head-hunter lured him away from NEC, Japan's top chip maker:

> Mr. Nakamura, head of our semiconductor division, asked me not to leave and even offered me a lateral move to any position within the company. He was so opposed to my leaving that Jerry Sanders [AMD's chairman] personally met with him to ask that I be allowed to join AMD. NEC is a valued customer of AMD and Mr. Sanders is a good friend of Mr. Nakamura's, so it was very awkward for both parties. But finally, after many discussions, Mr. Nakamura finally agreed.

Unlike the United States, where employment is seen as a legal contract that can be terminated at any time, Japanese have traditionally viewed employment as a social contract; leaving an employer is very difficult to do. The emphasis in Japanese society on personal feelings and human relationships makes job-hopping extremely tough for Japanese who value personal "face" and family honor above career prospects and salary. Yet this attitude is also changing. Recently, a Ministry of Labor survey found that almost half of all young employees would be interested if offered a better job opportunity. These *shinjinrui* are changing the face of corporate Japan.

Hiring Foreign Researchers

As Japanese companies seek new sources of creativity, they are looking overseas for top-notch researchers and managers. Electronic makers such as Fujitsu, Mitsubishi, NEC, and NTT Corporation are training foreign engineers in Japan for eventual relocation back to their overseas branches; these people are generally assigned to corporate laboratories under two- to three-year contracts. In 1987, for example, Fujitsu began sending recruiting notices to MIT and twenty-five other universities, which netted five specialists from China, Ireland, South Korea, and Taiwan. NEC is hiring more foreign researchers in Japan, who are treated as regular employees and receive the same benefits and status as Japanese employees. Even small computer graphics maker GRAPHICA has twenty foreign-born staff members. According to the Justice Ministry's Immigration Bureau, there were 54,736 foreign nationals with working visas in 1986. Of this increasing number of foreigners being hired by Japanese companies, however, only a handful are finding their way into research laboratories.

Foreign researchers face numerous obstacles. The language barrier, long working hours, endless meetings, and reliance on *nemawashi*—the time-consuming process of decision-making by consensus—can be frustrating for people accustomed to more independence, better wages, and privacy at work. As a popular saying has it: "There are no secrets in a Japanese office." Many foreigners burn out quickly and leave Japan because of stress, low pay, and lack of freedom. Nevertheless, more and more foreign researchers are being attracted to Japanese companies because of the opportunity to work on leading-edge technologies not actively being pursued in the West, such as computerized language translation and optocomputers.

Liberalized Corporate R&D Policies

Japanese companies are "pressure cookers" known for pushing their researchers to punishing limits of endurance in order to get their products to market before the window of opportunity closes. It is not uncommon for Japanese researchers to work 70–90-hour weeks to meet product deadlines. Corporate R&D heads themselves are under incredible pressure to match the latest developments of their competitors. For example, when Matsushita Electric introduced the first home bread-baking machine, Akimi Kamiya, managing director of Mitsubishi Electric, was immediately bombarded by requests from his distributors for a similar product.

Dr. Michiyuki Uenohara, who spent ten years at Bell Laboratories from 1957 and is now manager of NEC's Central Research Laboratory, observes, "We've been so busy catching up with the U.S. and Europe that we haven't had much time to carefully nurture our own proprietary technologies." When asked about the biggest change in NEC research policies during the last five years, he chuckles with a mixture of embarrassment and pride:

> In the past, creative people at NEC who failed totally lost face. It was difficult for them to start over. I'm trying to change that by creating an environment where people can fail several times and still come back. I tell young people it's okay to fail as long as they're trying hard and not playing it safe. It's difficult, especially for the older managers. But if we want to become more creative, we have to try . . . and we've had our share of failures and mistakes.

Under Dr. Uenohara, NEC's Central Research Laboratory has instituted a research policy that encourages both individual originality and group creativity. Individual researchers are allowed to leave group projects and go off on their own to focus on research topics of personal interest. By balancing the tensions between the individual and the group, Dr. Uenohara is trying to leverage both types of creativity. During his stay in the United States, he picked up another trick: weekly beer parties, à la Tandem Computer, in NEC's Sofa Plaza, where researchers can be heard laughing, arguing over new ideas, and telling jokes. By breaking up the seriousness of lab research, Dr. Uenohara is trying to inject some humor and creativity into his researchers' work.

Spinouts and Buyouts

Since 1985, Japanese companies have gone on a shopping spree buying land, buildings, companies, and now, technologies to take advantage of the huge "windfall profits" caused by the rapid devaluation of the U.S. dollar and other currencies. The more spectacular purchases grab the media's attention—the acquisitions of prime high-rises in New York, Los Angeles, and Chicago; equity investments in major securities houses on Wall Street; and Fujitsu's attempted buyout of Fairchild Semiconductor.

Stung by the strong anti-Japanese sentiment worldwide, many Japanese companies have maintained a low profile to avoid attracting media attention. They are investing small amounts of money—under $30 million—in promising high-tech companies, either in the form of equity ownership or acquisitions. Japanese companies—especially heavy industries—are investing in or acquiring leading-edge semiconductor, superminicomputer, software, and material companies (see Table 13–7). It is through low-level investments and acquisitions such as these, not the spectacular buyout, that some Japanese companies are obtaining valuable new technologies.

Entering Strategic Alliances

Since World War II, Japanese companies have primarily acquired technology through reverse engineering and licensing. Now, other countries' tougher enforcement of intellectual property laws and fear of the "boomerang effect"—being inundated by cheap look-alike products—have resulted in an infor-

TABLE 13–7
Japanese Investments and Acquisitions of High-Tech Companies

Japanese Company	Target Company	Amount ($ millions)	Product
Sanyo Electric	Icon Systems and Software[a]	5.6	Microcomputers
Kubota	MIPS Computers	25.0	Minisupercomputers
Kubota	Ardent Computer	n/a	Minisupercomputers
Kyocera/Mitsui	Chips and Technologies	1.5	Chip sets
Mitsubishi	Siltec[a]	35.0	Polysilicon
Nippon Kokan	Great Western[a]	n/a	Polysilicon
Rohm/Exar	Excel[a]	15.0	Memory chips
Canon	Next, Inc.	100.0	Personal computers

[a]Acquisitions
SOURCE: Dataquest, Inc.

mal "technology boycott" on Japan. Increasingly, Japanese researchers are no longer welcome in Western laboratories. Many U.S. technical conferences are closed to them. But one way of circumventing these barriers is to trade technology with a strategic partner.

The semiconductor industry is an example of this shift toward alliances as a competitive tool. Japanese semiconductor alliances are a relatively new phenomena. In the early 1980s these agreements generally involved one-way licensing, second-sourcing of U.S. microprocessors and semicustom chips, and joint development between U.S. and Japanese semiconductor equipment makers. Unwittingly, many U.S. companies gave away their technologies for up-front fees and royalties and failed to negotiate reciprocal exchanges of technology. As a result, several agreements ended up in court during the 1985 downturn as the Japanese partners began competing for the same market. Recently, Japanese alliances have become more sophisticated and varied, reflecting the shift toward higher value-added products. In 1987 over 140 alliances were signed, up sharply from previous years.

The foremost factor in Japan's new quest for strategic alliances is growing trade protectionism in the West. Since the U.S.-Japan Semiconductor Agreement, Japanese companies have become wary of depending on open markets. Overnight, Fujitsu, Oki Electric, and others were unable to sell DRAM chips to the United States at the high foreign market value (FMV) assigned to them by the Department of Commerce. Now they are teaming up with U.S. and European companies to secure offshore sales, chip design expertise, and distribution channels.

Tougher enforcement of intellectual property and patent protection laws is also driving Japanese semiconductor makers into alliances. The NEC/Intel suit, for example, was instrumental in the creation of Japan's TRON (The Real Operating Nucleus) Project to develop proprietary 32-bit microprocessors. Today, many Japanese chip makers are joining hands with other companies to exchange semiconductor know-how. The recent agreement between Hitachi and Mitsubishi to develop chip sets for HDTV and the Toshiba-Motorola manufacturing joint venture are cases in point.

Finally, the costs of developing a full line of products is so prohibitive that even large, vertically integrated Japanese companies cannot afford to go it alone. Building a new semiconductor plant today, for example, costs $150–250 million, and developing a standard 32-bit microprocessor can easily exceed $100 million, not including the software support required. With the entry of South Korea, Taiwan, Hong Kong, Singapore, China, and other Asian nations, semiconductor markets have become extremely competitive. To survive in these global markets, Japanese companies are seeking ways to accelerate their chip development. Strategic alliances are a way to leverage research capabilities by pooling them with those of a partner.

Toshiba is an example of a Japanese chip maker that has astutely used strategic alliances to its advantage. Since 1985, Toshiba has publicly announced twenty-four semiconductor alliances. During the 1985–86 industry downturn, Toshiba managed to gain market share and become number two in the chip industry because of its balanced product portfolio. It is also a well-known fact that Toshiba has dozens of supplier-user ties with Apple, Digital Equipment, and other foreign computer makers. During the "Toshiba-bashing" incident in Washington, D.C., these users were quickly mobilized in behalf of Toshiba to prevent Congress from enacting sanctions that would threaten the supply of critical components and subsystems from Toshiba.

Making Donations to Foreign Universities

Finally, Japanese companies are donating more money to foreign universities—as a goodwill gesture that could improve their corporate images and to gain access to leading-edge technologies. Japanese companies have endowed thirteen professorial chairs at MIT alone since the mid-1970s (see Table 13–8). MIT's Media Lab receives 18 percent of its funding from fourteen Japanese companies. And MIT's Industrial Liaison Program receives funding from forty-two Japanese companies. MIT is not unique. The California Institute of Technology has received funding from eleven Japanese companies; Columbia has more than ten Japanese donors. Other top schools receiving Japanese corporate donations include Cornell, New York University, Princeton, and Pennsylvania State University.

As Japanese companies globalize, their donations to universities in the United States, Europe, and Asia will grow. While most universities are still relatively open, strong political pressures—from local companies and national leaders—will close

TABLE 13–8
MIT Chairs Sponsored by Japanese Corporations

Year	Company	Research Topic of Chair
1974	Mitsui Group	Contemporary technology issues
1975	Japan Steel Federation	Metallurgy
1976	Matsushita Electric	Medical electronics
1979	Mitsubishi Group	International business management
1980	Mitsui Group	Contemporary technologies
1981	NEC	Computers and communication
1981	Toyota	Materials processing
1982	NEC	Software
1983	Kokusai Denshin Denwa	Telecommunications
1984	Kyocera	Ceramics
1984	TDK	Materials science
1987	Fukushiji Shoten	Education
1987	Nomura Securities	Finance

SOURCE: *Japan Industrial Journal* (27 January 1988).

many of their programs in the future. Universities will become extensions of local corporations as global competition heats up. Eastman Kodak's pressure on Rochester University to refuse business school admission to employees of rival Fuji Photo Film is a case in point. As universities seek closer ties with industry, Japanese donations raise critical issues that must be addressed: the potential conflict between public education and private enterprise, and academic freedom versus industrial secrets.

14

Whither Japan in the Twenty-first Century?

> The creativity of the Japanese people will be called into question from the latter half of the 1980s through the 1990s. The whole nation must work like one possessed to meet this great challenge.
>
> —Takuma Yamamoto
> President of Fujitsu

AS we enter the 1990s, Japan is undergoing a dramatic transformation that promises to shake the foundations of our global economy. It is becoming a fountainhead of new ideas and a source of emerging technologies and next-generation industries. Once manufacturers of imitative, low-cost products, Japanese companies have mastered the art of product innovation. Now they are moving ahead and forging their own distinctive style of creativity—a subtle blend of Japanese-style group creativity and Western-style individual creativity. The results are powerful. In field after field, from high-definition television to automobiles, they are poised at the leading edge.

Perhaps it was inevitable that the United States would be challenged in its final bastion of strength—science and technology. Since World War II, Japan has studied under the master and become a master in its own right. The proverbial caterpillar is turning into a butterfly. As its wings unfold, one wonders what its colors will be.

Japan is escaping from the shadows of Western culture and asserting its own cultural identity. A new era has begun: will we see the blossoming of Japanese creativity in the arts and sciences? Perhaps, if the Japanese can channel the enormous changes reshaping their nation. What are these changes?

Social Transformations

Probably the biggest change is occurring among the young people, who are embracing new values and lifestyles. Rejecting pressures to conform, to be housewives and corporate "salarymen," they are seeking jobs that offer more individual expression and freedom. In surveys, almost half of Japan's young people have shown interest in job-hopping. Many of them have traveled abroad or have been educated in the West. Others are discontented with the status quo. Although still a minority, their numbers are growing, and by the turn of the century they will become a tidal wave. They are already forcing changes in management practices. As the "baby boom" generation reaches middle age, the competition for scarce management positions is triggering an exodus of talented people from Japanese companies. Companies will stagnate unless they can motivate young people to stay. In the meantime, they are seeking outlets for their aspirations and dreams.

Working women are a major factor in the Japanese economy. According to the Women's Bureau of the Ministry of Labor, the 15.8 million working women accounted for 36 percent of Japan's work force in 1986, up from 32 percent in 1975. They represented 70 percent of the part-timers in 1986, versus 17 percent in 1975. As Japan expands its domestic economy to reduce trade friction, working women will be critical to the success of Japanese business because of their insights into local market needs. The extremely popular bread-baking machine, for example, was invented by teams of professional women at Matsushita Electric and Funai Electric. Many Japanese companies are hiring women as software programmers at satellite offices to tap local markets. Consumer surveys show that Japanese women make almost all of the home and automobile purchasing decisions. Through bottom-up decisionmaking processes, they are becoming influential in the choice of office automation equipment and software.

Japan's new wealth is creating opportunities for companies that sell luxury items and services, such as gourmet foods, fine wines, designer clothes, overseas tours, and sophisticated stereo equipment. The newly rich can afford to buy custom-made goods and expensive imports, a trend that is forcing Japanese companies to develop more creative products. In the past, quality sold. Today, the emphasis is on uniqueness and individuality. "High-sense"—the creation of products with subtlety and sophistication—is becoming a new consumer ethic. Mazda, for example, is a pioneer in the emerging field of "sensory" engineering: the development of new products that appeal to the five human senses.

Japan's rapidly aging population is worrying government policymakers and corporate leaders, who are seeking creative solutions to growing concerns for the health, safety, and personal well-being of the elderly. Japan will face the "silver shock" sooner than other industrialized countries. It already has the longest life expectancy in the world. In 1987 Japanese women lived an average of 81.4 years, while Japanese men lived 75 years. In 1987, 10.3 percent of the population was over 65 years of age; by 2020 that figure will be 24.6 percent, giving Japan the oldest population in the world. Pension and retirement systems, health care, housing, and social programs are already being strained. The incidence of cancer, strokes, osteoporosis, and other age-related diseases is rising. According to the Ministry of Health and Welfare, suicide is a growing problem among elderly people living alone. On the brighter side, the "silver society" is demanding new products and services, known as "silver business." Home medical care, fitness equipment, dietary foods, cultural activities, redesigned homes, modified audiovideo systems, and leisure travel are hot new fields of business.

Finally, Japan is not immune to unexpected political, economic, and natural "shocks." The Great Kanto earthquake in 1923, the "Nixon shock" of 1972 (when he visited China without consulting Japan), the "oil shocks" of 1973 and 1979, the Minamata mercury poisoning case and other environmental disasters during the 1960s and 1970s, and the "yen shock" of 1985—all revealed Japan's vulnerabilities. Because of the high concentration of people and wealth in Tokyo, for example, a major earthquake would derail the Japanese economy overnight and cause severe repercussions in global financial markets and trade relations. Of course, none of these shocks can be foreseen, but they can be anticipated through careful planning and preparation. How Japan anticipates and responds to these shocks will be a genuine test of its creative problem-solving abilities. Indeed, its role as a global financial power requires that it share the heavy responsibilities of ensuring stable economic growth and reciprocal trade worldwide.

The Barriers to Japanese Creativity

Overcoming the barriers to change, however, will be difficult for Japan. In this strongly group-oriented society, where individualism is still viewed with suspicion, becoming creative is easier said than done. Being individualistic often means facing ostracism, ridicule, and bullying.

Perhaps the biggest obstacle to change is the educational system, which lacks the flexibility to encourage individual creativity. Despite glowing reviews by many foreign observers, Japanese schools reward conformity and obedience more than originality and initiative. Creativity is given lip service, but the harsh reality of frequent examinations, cram schools (*juku*), all-night study marathons, high school rankings by test scores, and high-pressure "education mothers" (*kyoiku mama*) all too often nip it in the bud.

In the 1970s this author taught in the Okayama City school system, which provides an excellent, well-rounded education in math, science, social studies, the humanities, and the arts, especially at the elementary and early junior high school levels. Seventh-grade students were perky and full of curiosity and questions, but by the ninth grade they had become visibly exhausted from late-night cramming. They were much less inquisitive and asked few questions. High schools were so oriented toward college entrance examinations that almost all independent thinking and personal discovery came to a complete halt.

During the 1980s, Japanese schools have become even more high-pressured and competitive, even at the kindergarten level. Teachers must adhere closely to the prescribed textbooks. The norm is rote memorization and not disrupting the class by asking questions or engaging in long discussions, in order to get through the voluminous amounts of assigned curriculum. And the infamous "exam hell" during high school is exactly that—a student's nightmare. As a result, Japanese universities are seen as four-year resorts for resting up until one's grueling career begins.

Many corporate managers frankly admit that young people must unlearn many bad habits picked up at school, such as blindly following orders and lack of initiative and curiosity. While this educational approach has worked in the past, it is counterproductive to Japan's quest for greater creativity in its research laboratories and service industries. MOE is investigating educational reform, but the pace of change will be glacial among this elite group of bureaucrats and reformers.

Sexual discrimination is a serious obstacle to creativity in Japan. In many respects, Japan is working with only "half a brain" because of job discrimination against women. Despite passage of the Equal Job Opportunity Law of 1986, Japanese companies regularly overlook women in making key hiring, transfer, and promotion decisions. At the end of 1988, only 7.6 percent of all Japanese companies had adopted equal employment policies. To most senior managers, working women are still viewed as tea-servers or part-timers making a little pin money before marriage or after raising their children. The hiring of college-educated women has doubled from 36 percent of new graduate hires before passage of the new law to 79 percent now. But the prospects for women trained in math, science, and engineering are dismal. Moreover, female high school students are routinely discouraged from enrolling in "male" courses. As a result, most Japanese women end up in conventional service-oriented careers, such as nursing, banking, teaching, advertising, fashion design, and retailing. While the influx of women professionals has enhanced creativity in these sectors, the traditional male-dominated professions, such as engineering, research, and management, still remain overwhelmingly male. Companies and individuals in these professions often exhibit the least creativity because of their closed, hierarchical nature. Opening up to women professionals would infuse these professions with fresh ideas and energy, but the short-term prospects for that happening remain poor. Japanese women are likely to find better promotion opportunities in start-up companies or foreign subsidiaries.

Japan's overweening self-confidence and pride are becoming a major hindrance to creativity. While the Japanese can be justifiably proud of their postwar accomplishments, opinion leaders are becoming noticeably more arrogant and disdainful of foreigners. Ex-Prime Minister Yasuhiro Nakasone's derogatory comments about the "intellectual inferiority" of American blacks and Hispanics is a case in point. Most revealing was the absence of public outcry over his comments; many Japanese privately believe that Nakasone was only stating the obvious. This condescending attitude toward foreigners is not confined to Japan's leadership; it is commonplace throughout the country. People returning from abroad, for example, are often viewed as "contaminated" by foreign influence. Asian and African students are not welcome in most places. Trade friction is blamed on foreign nations, even though the causes are complex and multilateral. The "Japan as Number One" mentality could ultimately undermine the Japanese economy.

Rather than see Japan become a global leader, we could witness instead the "closing of the Japanese mind." Not only would the world lose out, but Japan would run the risk of isolating itself from the new ideas and technologies it needs to reach first rank in basic research and innovation.

The Emerging Challenge

Despite these many challenges, we are likely to see Japan becoming more creative in the 1990s. The transformation will not occur suddenly, nor command our attention at first. We will be preoccupied with long-term problems—regaining our international competitiveness and shoring up our crumbling infrastructure and educational system. Toxic wastes, drugs, street crime, assault weapons, bank bailouts, and trade and budget deficits will absorb our attention at home. Training an unprepared work force will drain additional resources. *Perestroika,* China, and Europe's market opening in 1992 will divert our attention. Frustrated with cultural barriers and slow progress in the Japanese market, many Western companies will seek quick profits elsewhere. They will be enticed by Asia's booming economies, but will be hobbled by distance and a lack of marketing savvy.

Meanwhile, Japan will be strengthening its scientific infrastructure. Japanese corporations will be forced to modify their management policies to accommodate the growing numbers of working women, independent-minded young people, aging managers, and returnees from abroad. Change will come slowly but surely within classrooms, government ministries, and the corporate suite. New technologies will come pouring out of the new corporate laboratories and software development centers scattered throughout Japan. Many products "created in Japan" will not be available in the West. They will be either designed for domestic use or shipped to other markets. Products sold in the West will increasingly be designed and manufactured by local Japanese subsidiaries.

The signs are all around us. Japan has the money, the talent, the technologies, and the will. Japanese companies have already passed us by in optoelectronics, high-definition television, memory chips, robotics, factory automation, and computerized language translation. They are now making a bid for superiority in supercomputers, fine ceramics, biotechnology, aerospace, superconductors, nanomechanism technology, medical electronics, marine development, optical communications, and automotive electronics. "Created in Japan" may soon replace "Made in Japan" as a symbol of industrial excellence.

By the mid-1990s Japan's technological prowess could overwhelm the West, and the political shock waves of losing one next-generation industry after another to Japan will be severe. By the late 1990s Japanese companies will have mastered the entire "mandala of creativity." They will excel not only at refining and recycling ideas but also at exploring and generating new ideas. While we stubbornly cling to the romantic but outdated notion of "rugged individualism," we may find ourselves outgunned by Japan's potent new brand of creativity. By the year 2000 we may be left behind unless we wake up and reconsider our notions of creativity.

What new technologies can we expect to see from Japan during the 1990s? In 1988 *Nikkei Electronics* magazine polled twenty-three top Japanese research managers about new electronic products they expect to commercialize. Their answers provide startling insights into the future. They feel confident that the following products will be "created in Japan":

- The fusion of HDTVs and computers into videocomputers
- Wall-size displays for HDTV and computers
- Intelligent copiers, printers, and other office automation equipment
- Video, image, text, data, and voice messages transmitted over digital networks
- Home service robots capable of cleaning, sweeping, and security
- Car navigation systems using sensors and microprocessors
- Factories and cars run by optical systems
- Superconducting trains, chip-making equipment, and earthquake detectors
- Gigabit memory chips capable of storing dozens of books
- Proprietary 32-, 48-, and 64-bit TRON microprocessors for use in home, consumer, office, and factory automation
- Supercomputers capable of one TFLOP (one trillion floating point operations per second)
- Powerful AI chips and software for medical systems,

chip design, and natural language translation computers and telephones
- Fuzzy logic controllers for factory automation and subway operations

If these forecasts are unsettling, an even more bullish technology forecast was issued in 1987 by STA, which outlined Japan's technology strategy from 1987 to the year 2015. The findings were the result of a survey in which experts from government, industry, and academia gave their best estimates about future breakthroughs from Japan. To date, STA has issued four technology forecasts with a track record of 70 percent accuracy. Here is a partial listing of Japan's technology targets for the next twenty years:

1994—Computer-aided design of computer chips with more than one million gates
1996—New protocol technology to link communications networks
1997—Sludge removal from ocean floor; fishing site cleanup; underwater robots used to 1,000 feet
1999—AI in aircraft management and control
—Nursing robots to help the elderly and handicapped
—Local disaster forecasting and prevention systems
2000—Electronic data storage safe from human errors, viruses, or natural disasters
2001—Artificial organs not rejected by the human body
2002—Prevention of cancer cell spread in the body
—Quick turnaround of bug-free, complex software
—Space robots with sophisticated artificial intelligence
2003—International digital communications network
—High-performance materials for space travel
—Medicine and semiconductor production in space
2005—Turn cancer cells back to normal cells
2011—Room-temperature superconductors for industrial machinery

If Japanese companies achieve one-fourth of these goals, global industries will undergo a major restructuring. Indeed, whole industries will be born and destroyed. The biggest challenge from Japan will not be technical, but psychological and social. Just as the computer industry is shifting its focus from hardware to software and "groupware," we will see the rise of a Japanese "humanware" industry—the creation of new products and services that contribute to greater social and physical well-being. (Current humanware concepts include time-sharing, time-shifting, job-sharing, and telecommuting.) Humanware engineering could well become a major growth industry in the twenty-first century, on the scale of the Industrial Revolution. A global creativity revolution would be its precursor.

Techno-Nationalism or Techno-Globalism?

Japan is at a historic crossroads. With the passing of Emperor Hirohito, its link to the past has been severed. Now an economic and technological superpower, Japan is being courted as a major shaper of decisions and events. In many circles it is seen as the world's banker and provider of jobs. Japanese companies wield enormous clout with decisionmakers worldwide, a situation that is creating new opportunities and tensions. Japan's expansionism is coming under closer scrutiny. There is a growing sentiment around the world that Japan should show more creative leadership—not just creative mercantilism, but creative science, education, diplomacy, and philanthropy. Japan is being pressed to come out from behind its island mentality and its wealth and demonstrate genuine leadership in solving global problems.

The Japanese seem uncertain, and almost unwilling, about their new role because of the risks involved. They are weighing the burdens of global leadership. As Japan pursues creativity, will the gap between its rich and poor widen? Will its upscale products and services exacerbate "Japan bashing"? Will other nations use threats of protectionism to force Japan to share its technology? As Japanese companies leapfrog their Western competitors, will they be slapped with more tariffs, quotas, and export controls? Already, U.S. universities are discouraged from buying Japanese supercomputers, and the Pentagon is pressuring Japan to classify more commercial research and tighten its export controls. Will national security be used as an argument to blunt Japan's nascent scientific prowess?

Whether it shares or hoards technologies, Japan will be a major player sharply criticized on all sides. What can it do in this no-win situation?

One test of Japan's vow of internationalism is its patent system, which is being challenged in a long-standing dispute between Mitsubishi Electric and Fusion Systems Corporation in the United States. The case began in 1985, when Fusion discovered that Mitsubishi had filed more than 160 patents in Japan for products similar to its linear microwave-actuated ultraviolet lamp. Fusion had been selling the lamps in Japan since 1975. In its defense, Mitsubishi argued that its spherical lamps did not infringe on Fusion's patent because they were significantly different in design. Fusion countered that Mitsubishi had made only "trivial changes" to its design.

The real issue lies in the difference between the Japanese and U.S. patent systems. While the U.S. system awards patents to the first to invent, the Japanese system awards patents to the first to file. Sosuke Sato, deputy director of the international affairs division of the Japanese Patent Office, defends Japan's approach by arguing that the U.S. system protects individuals, while the Japanese system balances individual rights with broader social and industrial interests.

This argument may appear self-serving to Westerners. What if Japanese companies hold key patents for cures to fatal diseases? Shouldn't they be required, as Sosuke Sato suggests, to relinquish their rights for the benefit of non-Japanese? If not, is Japan adhering to a double standard? Japan's patent system also raises serious questions about the primacy of individual rights. To what extent do intellectual property rights supersede social interests? Should discoverers of AIDS treatments, for example, be given exclusive patent rights, to the possible detriment of AIDS victims? As biotechnology becomes a dominant industry in the twenty-first century, would public health be better served under individual patent rights or under a new form of collective ownership? If teams of researchers around the world contribute to a breakthrough, should lawsuits be allowed to block the diffusion of the new technology to beneficiaries? In Western courts, individual inventors usually have the upper hand. How can this custom be reconciled with legal decisions in more group-oriented cultures?

The Mitsubishi/Fusion case has reached a stalemate, and it is unlikely the Japanese Patent Office will budge from its advantageous position, which has nevertheless raised doubts about Japan's commitment to promoting creative research. As long as it is easier to copy under Japanese law, Japanese companies will continue to do so. Under the new U.S. "Super 301" trade law, foreign companies such as Mitsubishi can be prevented from importing products into the United States if an American company believes they violated its patent rights. The burden of proof will fall on Japanese companies to demonstrate that they did not copy.

Like other Asian nations, Japan is considering new intellectual property laws to protect the rights of its inventors. In the early 1980s MITI and MOE's Cultural Agency had an ongoing debate over whether software was covered by copyright or patent protection laws. Ultimately, international copyright law prevailed, but MITI is still considering how to expand patent laws to serve the interests of Japanese industry.

Access to research laboratories will be another litmus test for Japan in the 1990s. Currently, Europe and the United States are closing their laboratories to Japanese researchers because of lack of access to Japanese laboratories. The Japanese government is gradually opening its laboratories to foreign researchers. The real issue, however, concerns the openness of Japanese corporate laboratories, where most of the crucial commercial research is conducted. Although individual companies are beginning to exchange researchers, foreigners still stand to lose because of the language barrier and the close-knit nature of Japanese corporate culture.

Over time, these obstacles will lessen as government ministries open their doors and Japanese corporations globalize their operations. But change will not come quickly. The Japanese have been much too successful to make basic changes in their industrial and corporate policies. Indeed, Japan's economic success could ultimately become its Achilles' heel. We have seen how economic supremacy crippled England, and how it has toppled many U.S. industries. If Japan stumbles, it will not be caused by industrial, educational, or financial weaknesses. It will happen because of what John Kenneth Galbraith once called "the arrogance of power." It will happen because of Japanese insensitivity to the powerful political, economic, and environmental forces reshaping the globe. Avoiding this, in the long run, will be the true test of Japan's creativity.

Whither Japan?

BIBLIOGRAPHY

Chapter 1. The Sound of One Hand Clapping

Broad, William J. "Novel Technique Shows Japanese Outpace Americans in Innovation." *New York Times,* 7 March 1988.

"U.S. Still Stumbling in Technology Race: Firms Not Turning Patents into Products." *Manufacturing Week* (18 April 1988): 22.

Morita, Akio. *Made in Japan.* New York: E. P. Dutton, 1986.

Narin, Francis, and Dominic Olivastro. "Identifying Areas of Leading Edge Japanese Science and Technology—Activity Analysis Using SIC Categories and Scientific Subfields." First Interim Report, National Science Foundation Grant No. SRS-8507306, 19 May 1986.

Rudolph, Barbara; Yukinori Ishikawa; and Thomas McCarroll. "Eyes on the Prize: Japan Challenges America's Reputation for Creativity and Innovation." *Time* (21 March 1988): 50–51.

Wysocki, Bernard, Jr. "The Last Frontier: Japan Assaults the Last U.S. Bastion—Its Lead in Innovation." *Wall Street Journal,* 14 November 1988, sect. 4 supplement.

Chapter 2. East Meets West: The Yin and Yang of Creativity

De Bono, Peter. *Lateral Thinking.* London: Penguin Books, 1988.

Drucker, Peter F. *Innovation and Entrepreneurship: Practice and Principles.* New York: Perennial Library, 1985.

Kikuchi, Makoto. *Japanese Electronics: A Worm's-Eye View of Its Evolution.* Tokyo: Simul Press, 1983.

May, Rollo. *The Courage to Create.* New York: Bantam Books, 1983.

Ray, Michael, and Rochelle Myers. *Creativity in Business.* Garden City, N.Y.: Doubleday, 1986.

Riggs, Henry E. "Innovation: A U.S.-Japan Perspective." Paper for the High Technology Research Project at Stanford University, March 1983.

Rosenfeld, Robert, and Jennry C. Servo. "Business and Creativity: Making Ideas Connect." *Futurist* (August 1984): 20–25.

Tatsuno, Sheridan M. "Fujitsu America: An Interview with Katsuhide Hirai." San Jose, California, 30 March 1988 (unpublished).

Von Oech, Roger. *A Whack on the Side of the Head.* New York: Warner Books, 1983.

———. *A Kick in the Seat of the Pants.* New York: Perennial Library, 1986.

Chapter 3. The Creative Samurai: Japanese Intrapreneurs and Small Businesses

Aoyama, Shuji, and S. K. Subramanian. "Restructuring Mature Businesses—Asian Forum Looks at Japanese Trends." *Journal of Japanese Trade and Industry,* no. 4 (1988): 45–48.

Drucker, Peter F. *Innovation and Entrepreneurship: Practice and Principles.* New York: Perennial Library, 1985.
Fukukawa, Shinji. "MITI's Policies for Fiscal 1987: A Policy Plan for Restructuring Japanese Industry." *Journal of Japanese Trade and Industry,* no. 1 (1987): 47–50.
Hasegawa, Keitaro. *Japanese-Style Management: An Insider's Analysis.* Tokyo: Kodansha International, 1988.
Higurashi, Ryoichi. "Sanrio: Cute Products Spell Success." *Journal of Japanese Trade and Industry,* no. 6 (1987): 30–31.
Kanamori, Hisao. *Innovation and Industrial Structure.* Tokyo: Japan Economic Journal Press, 1987.
Kishimoto, Yoriko. "Changes in Japanese Venture Capital." *Venture Japan* 1 (1988): 20.
"Japan's 1987 Hit Phenomenology." *Look Japan* (February 1988): 25.
Maeda, Shozo. "Finding a Niche for a Novelty." *Journal of Japanese Trade and Industry,* no. 3 (1988): 18–19.
Matsumura, Moritaka, ed. *The Best of Japan—Innovations: Present and Future.* Tokyo: Kodansha International, 1987.
Ministry of International Trade and Industry (MITI). *New Business Information.* Tokyo: MITI, 10 October 1987.
———, Small and Medium Enterprise Agency. *White Paper on Small and Medium Enterprises in Japan.* Tokyo: MITI, 1987.
———, Small and Medium Enterprise Agency. *Venture Business Prospects and Issues: Midterm Report.* Tokyo: MITI, 1984.
Misawa, Mitsuru. "New Japanese-Style Management in a Changing Era." *Columbia Journal of World Business* (Winter 1987): 13.
Morishima, Michio. *Why Has Japan Succeeded?: Western Technology and the Japanese Ethos.* Cambridge, England: Cambridge University Press, 1982.
Murakami, Morio. "Japanese-Style Management Threatened." *Journal of Japanese Trade and Industry,* no. 2 (1987): 28–30.
Naito, Yosuke. "Ongoing Transformation in Japanese Management." *Journal of Japanese Trade and Industry,* no. 4 (1986): 46–49.
"Market Creation." Series of 135 articles in *Nikkei Sangyo Shimbun* (Japan Industrial Journal) (November 1986–March 1987).
"Small Businesses Riding Current Trends: The Nikkei Small Business 1000 Survey." *Nikkei Sangyo Shimbun* (Japan Industrial Journal), 16 October 1987, sect. 2.
Nonaka, Ikujiro. "Focusing on Automatic Technology—Company Strategy: Minolta Camera." *Look Japan* (May 1987): 16–19.
———. "The Soapmaker's Floppy Discs." *Look Japan* (April 1987): 18–19.
Ohmae, Kenichi. *The Mind of the Strategist.* New York: McGraw-Hill, 1983.
Okumura, Akihiro. "Restless Creativity—Company Strategies: Kyocera." *Look Japan* (November 1987): 16–17.
———. "New Markets, New Tricks—Company Strategies: Matsushita Electric Industrial Co. Ltd." *Look Japan* (December 1987): 18.
Okunuma, Ariyoshi. "Japan's Changing Economic Structure." *Journal of Japanese Trade and Industry,* no. 5 (1987): 10–13.
Ozawa, Kiyoshi, ed. *New Frontier Enterprises.* Tokyo: Toyo Economic Press, 1983.
"Watch These Next-Generation Companies." *Shuukan Diamond* (2 May 1987): 18–36.
"New Products, New Technologies, and New Materials." *Shuukan Diamond* (4 July 1987): 62–76.
"Total Revolution: The New Business 200." *Shuukan Diamond* (8 August 1987): 62–77.
"Eighty-eight Services for the Home." *Shuukan Diamond* (22 August 1987): 70–75.
"The New Business 180 Greet Good Times." *Shuukan Diamond* (19 December 1987): 87–98.

Sumino, Masafumi. "Venture Businesses in the New Technology Era." Kamakura: Nomura Research Institute, 1983.
Takatsu, Akira. "MITI's Fiscal 1988 Small- and Medium-Size Business Policies: Fusion Promotion Measures." *Electronics* (Electronics Industry Association of Japan) 28 (March 1988): 14–21.

Chapter 4. Historical Origins of Japanese Creativity

Ando, Shoei. *American Transcendentalism and Zen.* Tokyo: Hokuseido Publishing, 1969.
———. *American Literature and Zen.* Tokyo: Eihō, 1970.
Imaizumi, Hiroaki. *The Mandalart Technique.* Tokyo: Japan Jisseki Publishing, 1988.
Itogawa, Hideo. *Creative Idea Method.* Tokyo: President Publishing, 1984.
Kanba, Wataru. *Super-Idea Methods in the Information Era.* Tokyo: Keizaikai, 1986.
Kayano, Takeshi. *Creativity.* Tokyo: Mikasa Shobo, 1988.
Kuhn, Thomas. *The Structure of Scientific Revolutions.* Chicago: University of Chicago Press, 1962.
Kurabe, Yukio. *"Heavens No" Idea Method.* Tokyo: PHP Publishing, 1986.
Mori, Masahiro. *The Buddha in the Robot: A Robot Engineer's Thoughts on Science and Religion,* translated by Charles S. Terry. Tokyo: Kosei, 1974.
———. *Advice on Nonduality.* Tokyo: PHP Business Library, 1987.
Morishima, Michio. *Why Has Japan Succeeded?: Western Technology and the Japanese Ethos.* Cambridge, England: Cambridge University Press, 1982.
Moritani, Masanori. *Japanese Technology: Getting the Best for the Least.* Tokyo: Simul Press, 1982.
Morris, Ivan. *The World of the Shining Prince.* London: Penguin Books, 1964.
Nakayama, Sadakazu. *Zen and the Brain.* Tokyo: PHP Publishing, 1984.
Sansom, George. *A History of Japan.* Stanford, Cal.: Stanford University Press, 1958.
Sato, Hidenori. *Creative Idea Method: Improving Your Creativity.* Tokyo: Nippon Jitsugyo Shuppansha, 1988.
Shinagawa, Yoshiya. *The Brain and Creativity.* Tokyo: Daiwa Shobo, 1985.
Suzuki, Daisetz Teitaro. *Studies in Zen.* New York: Dell/Delta, 1955.
Tatsuno, Sheridan M. "Japan's Pursuit of Creativity." In *The Technopolis Strategy: Japan, High Technology, and the Control of the Twenty-first Century,* edited by Sheridan M. Tatsuno, pp. 215–218. New York: Brady/Prentice-Hall Press, 1986.
Von Oech, Roger. *A Kick in the Seat of the Pants.* New York: Perennial Library, 1986.

Chapter 5. Sairiyo: Recycling the Past

Kawabata, Yasunari. *The Snow Country.* New York: Alfred A. Knopf, 1957.
Kodama, Fumio. " 'Technology Fusion' Yields Innovations from Breakthroughs." *Japan Economic Journal* (13 August 1988): 22.
Lee, O-Young. *Smaller Is Better: Japan's Mastery of the Miniature.* Tokyo: Kodansha International, 1982.
Ministry of International Trade and Industry (MITI). "The Fundamental Outline of Twenty-first Century Industrial Society." May 1986.
Morita, Akio. *Made In Japan.* New York: E. P. Dutton, 1986.

Chapter 6. Tansaku: Exploring New Ideas

Unger, J. Marshall. *The Fifth Generation Fallacy.* New York: Oxford University Press, 1987.
Vogel, Ezra. *Japan as Number One.* New York: Harper & Row, 1980.

Chapter 7. Ikusei: Nurturing Creative Ideas

Choy, Jon. "Update on Japan's Venture Capital Industry: Problems and Promises." Japan Economic Institute (JEI) Report 29A. Washington, D.C.: JEI, 29 July 1988.
Kishimoto, Yoriko. "Venture Capital à la Japonaise." *Venture Japan* 1 (1988).
Kurata, Masashi. "Plus & Company: Stationery on the Move." *Journal of Japanese Trade and Industry,* no. 3 (1987): 39–40.
Makino, Noborn. *Decline and Prosperity: Corporate Innovation in Japan.* Tokyo: Kodansha International, 1987.
"Venture Capital Recovers—1987 Survey Results." *Nikkei Sangyo Shimbun* (Japan Industrial Journal) (23 February 1988): 7.
Ouchi, William. *The M-Form Society.* Reading, Mass.: Addison-Wesley, 1984.

Chapter 8. Hassoo: Generating Breakthroughs

Honda, Soichiro. *Each Day a New Perspective.* Tokyo: PHP Business Library, 1985.
Inouye, Shodo, and Shigeo Hatakeyama. *Introduction to the KJ Method.* Tokyo: Japan Management Association, November 1971.
Matsumura, Yasuo. *The Miracle of the MY Method.* Tokyo: Kodansha International, 1988.
Nakayama, Masakazu. *Promoting Creativity.* Tokyo: PHP Press, 1988.
Nishizawa, Junichi. *Idea Methods for Reading Ten Years into the Future.* Tokyo: Kodansha International, 1986.
Takahashi, Hiroshi. *The Creativity Handbook.* Tokyo: Japan Management Association, March 1988.
Takemura, Kenichi. *The Era of Genius.* Tokyo: Gakushu Kenkynsha, 1987.

Chapter 9. Kaizen: Refining Ideas

Lee, O-Young. *Smaller Is Better: Japan's Mastery of the Miniature.* Tokyo: Kodansha International, 1982.
Matsumura, Moritaka, ed. *The Best of Japan—Innovations: Present and Future.* Tokyo: Kodansha International, 1987.
Morita, Akio. *Made in Japan.* New York: E. P. Dutton, 1986.

Chapter 10. High-Definition Television: The Next-Generation Video Battlefield

Alster, Norm. "TV's High-Stakes, High-Tech Battle." *Fortune* (24 October 1988): 161–170.

Brody, Herb. "The Push for a Sharper Picture." *High Technology Business* (April 1988): 25–29.
Burgess, John. "Chances Are High TV Breakthrough Is in the Air." *Washington Post,* 10 October 1987.
Cohen, Andrea. "Look Sharp, Suppliers: HDTV Will Be a Great Show." *Electronic Business,* (1 April 1988): 28–29.
Doherty, Richard. "EIA Boosts HDTV: Out of Lab, into Home." *Electronic Engineering Times,* 5 December 1988.
Elkus, Richard. "The Impact of HDTV." Unpublished speech given to American Electronics Association HDTV Task Force. 1989.
Feibus, Michael. "On Japanese Dominance, HDTV, and Signetics: An Interview with Cees Koot of N. V. Philips." *San Jose Mercury-News,* 24 October 1988.
Free, John. "Sharpies: IDTV." *Popular Science* (November 1988): 54–62.
Izumikawa, Shinichi. "Moving Closer to High-Definition TV: Scanning Density Improves Picture Quality." *Asian Electronics Union* (September 1985): 97–100.
"High Resolution EDTV Attracting Great Attention." *Japan Electronic Engineering* (March 1987): 54–59.
Lammers, David. "Japan Hypes HDTV: Tokyo Steps on the Gas." *Electronic Engineering Times,* 14 March 1988.
Maeno, Kazuhisa. "Changing Picture in Video Kingdom." *Journal of Japanese Trade and Industry,* no. 3 (1988): 28–30.
Makino, Shinichi. "Development of High-Definition TV Systems in Present-Day Japan." *Japan Electronic Engineering* (March 1987): 28–30.
"Hi-Vision Starts Up." *Nikkei Electronics,* no. 457 (30 October 1987): 107–139.
Port, Otis. "HDTV: Washington Still Isn't Receiving the Signal." *Business Week* (21 November 1988): 114.
Sakashita, Yuko. "Clearing Up Old Problems: Media-Satellite Broadcasting." *Look Japan* (October 1987): 20–21.
Sakurai, Miyoko, and David Lammers. "Japan Reception: TV or Alphabet Soup?" *Electronic Engineering Times,* 11 July 1988.
Sazegari, Steve. "Does HDTV Offer an Opportunity for Revival of the U.S. Consumer Electronics Industry?" Dataquest Research Bulletin, TCIS Code 1988–34 (September 1988).
Udagawa, Hideo. "NHK: A Tale of Dreams." *Tokyo Business Today* (November 1988): 20–21.
Woolnough, Roger. "A New Era Ahead for European TV: A Megabit Boost in Quality." *Electronic Engineering Times,* 3 October 1988.

Chapter 11. The Computer Bazaar of the Future

Dambrot, Stuart. *Managing Automation* (February 1988): 38–40.
EDP Japan. "Sigma Prospectus Signals Software Development Push." Vol. 12 (24 December 1985).
Feigenbaum, Edward A., and Pamela McCorduck. *The Fifth Generation.* Reading, Mass.: Addison-Wesley, 1983.
Fernbach, Sidney, et al. "Supercomputers: How Long Can We Afford to Wait Before We Respond?" Report of the Institute for Electrical and Electronic Engineering's Scientific Supercomputer Subcommittee. *New Technology Week* (15 August 1988): 6–9.
Johnson, R. Colin. "Compiler Boosts Fuzzy Logic." *Electronic Engineering Times,* 5 December 1988.

Lammers, David. "Is ICOT Overambitious?" *Electronic Engineering Times,* 12 December 1988.

McCormack, Richard. "Uncertainty Clouds the Future of U.S. Optics Industries." *New Technology Week* (16 February 1988): 1–5.

Nano, Hiko. *The Neural Computer Revolution.* Tokyo: Kodansha Business, 1989.

National Academy of Engineering. "Photonics: Maintaining Competitiveness in the Information Age." 1988.

Pau, L. "AI in Japan." *Look Japan* (October 1988): 40.

Sakamura, Ken. *Computers Changed by TRON.* Tokyo: Nihon Jitsugyo Shuppansha, April 1987.

———. *New Concepts from the TRON Project.* Tokyo: Iwanami Shoten, February 1987.

Suzuki, Michio. "AI Applications in Electricity." *Look Japan* (November 1988): 21.

Unger, J. Marshall. *The Fifth Generation Fallacy.* New York: Oxford University Press, 1987.

U.S. Department of Commerce, National Technical Information Service. *Japanese Technology Evaluation Program (JTECH) Panel Report on Advanced Computing in Japan.* Prepared for the National Science Foundation by Marvin Denicoff, chairman (December 1987).

U.S. Senate, International Trade Commission. *U.S. Global Competitiveness: Optical Fibers, Technology and Equipment.* Report to the Committee on Finance. Investigation 332–333 (Publication 2054), 1988.

Wood, Robert Chapman. "Technology with Intuition: These Computers Will Understand Body Language." *High Technology Business* (December 1988): 10.

Chapter 12. Visions of Superconductors

Anderson, Anne. "Japan and Superconductivity: Everybody Participates." *New Technology Week* (22 February 1988): 1.

Burgert, Philip. "U.S. Stumbling in Superconductors." *Manufacturing Week* (4 April 1988): 22.

Cambridge Report on Superconductivity. "Commercial Superconductor Developments: The Future Is Now, and with the Japanese" (November 1987).

Cassidy, Robert. "Japan's Open Door: Is the U.S. Missing Its Golden Opportunity?" *Research and Development* (November 1988): 81–84.

Choy, Jon. "Superconductivity Projects Launched." Japan Economic Institute (JEI) Report 23B (17 June 1988).

———. "Superconductors: An Update on Japanese R&D." *Journal of the American Chamber of Commerce in Japan* (April 1988): 47–60.

Itakura, Kimie. "Creating New Metals: Materials Research." *Look Japan* (May 1987): 23.

Kimura, Hiroko. "Magnetically Levitated Train Shaping Up." *Asahi News* (10–11 November 1988): 4.

Ministry of International Trade and Industry (MITI). *Mid-term Report of the Industrial Superconductor Technology Development Roundtable, 24 August 1987* (unofficial translation). Reprinted in *Journal of the American Chamber of Commerce in Japan* (April 1988): 47–60.

Naito, Motoko. "Superconductivity: Shoji Tanaka—The Race Heats Up." *Look Japan* (June 1988): 24–25.

National Science Foundation. Report Memoranda nos. 122, 123, 128, 129, 138, 139, and 140—Visits to Japanese Superconductivity Research Laboratories, May 1987–September 1987.

"Are the Japanese Ahead? How Four Industrialists View Superconductivity." *New Technology Week* (18 April 1988): 14–16.

Overell, Edward. "Japan Embarks on $200 Million Superconducting Generator Program." *New Technology Week* (23 November 1987): 1.
———. "Japanese, Germans, French (and Not the U.S.) Battle to Supply U.S. High Speed Transport." *New Technology Week* (23 November 1987): 4–5.
———. "Japan's Maglev: Full Steam Ahead." *New Technology Week* (30 November 1987): 4–5.
———. "Japan's Maglev Companies Fine-tune Technologies to Capture Future Market." *New Technology Week* (7 March 1988): 3–12.
Torii, Hiroyuki. "JR Pushes Maglev Research in Race for Faster Train." *Japan Economic Journal* (16 July 1988): 1–5.
Yoder, Stephen Kreider. "Japan Is Racing to Commercialize New Superconductors." *Wall Street Journal,* 20 March 1987.

Chapter 13. Satori in the Laboratory: The Challenges Facing Japanese Researchers

Anderson, Alun. "Unused Researchers to Arms." *Nature* 309 (1984): 659.
Anderson, Anne. "Hitachi's R&D Budget Growing Year by Year." *New Technology Week* (29 February 1988): 2.
Arimoto, Takeo. "New World Science: Human Frontier Science Program." *Look Japan* (July 1988): 22–23.
Asai, Tsuneo. "Great Leaps Forward: High-Tech Internationalization." *Look Japan* (May 1988): 22–23.
Debevoise, Malcolm. "Japanese Scientist Overseas: Leo Esaki—Do What You Want to Do." *Look Japan* (March 1988): 26.
"Creative Scientific Technology Promotion Enterprise Chooses New Themes." *Dempa Shimbun* (4 June 1987): 41.
Esaki, Leo. *The Spirit of Creativity.* Tokyo: Mitatsu Shoten, November 1987.
Fuchs, Peter. "Exporting American Academia." *Business Tokyo* (May 1988): 51.
Hidaka, Satoshi. *The Geneology of Creativity.* Tokyo: Diamond Publishing, 1985.
High Tech and Management Research Association. *New Idea Methods of Technology Staff Offices.* Tokyo: Diamond Publishing, 1987.
Imai, Kenichi. "The Organization of Creative Innovation." *Economics Today* (Spring 1987): 36–61.
Inoue, Yuko. "Foreign Firms Set Up Research Beachheads." *Japan Economic Journal* (14 November 1987): 4.
———. "Japanese Firms Rapidly Increase Global Research and Development." *Japan Economic Journal* (24 September 1988): 7.
"Technological Thresholds for New Century." *Japan Economic Journal* (October 1987–February 1988).
"Japan to Open Its Research Doors to Ease Technological Friction with U.S." *Japan Economic Journal* (14 November 1987): 9.
"Twenty-first Century Outlook—Execs' Views Vary on Technology Future." *Japan Economic Journal* (15 October 1988): 11.
Johnstone, Bob. "Catalyst for Change: Multinational Chemical Firms Seek Foothold in Japan." *Far Eastern Economic Review* (17 December 1987): 112.
———. "Back to the Basics: Japan Begins Search for Excellence in Scientific Research." *Far Eastern Economic Review* (12 January 1989): 57.
Kawasaki, Masahiro. "Researching Policies." *Look Japan* (November 1988): 28.
Kojima, Takeshi. "Charting a New Course: Ishikawajima-Harima Heavy Industries." *Look Japan* (August 1988): 20.

Lee, Chris. "The Social Organization of Japanese Research Science: Restrictions and the Overdoctor." Student paper for Anthropology 155, Professor Thomas Rohlen, Stanford University, 12 May 1986 (unpublished).

Maurer, P. Reed. "Ready for an R&D Breakout." *Business Tokyo* (May 1988): 41–43.

Misawa, Mitsuru. "New Japanese-Style Management in a Changing Era." *Columbia Journal of World Business* (Winter 1987): 9–17.

Morishima, Michio. *Why Has Japan Succeeded?: Western Technology and the Japanese Ethos.* Cambridge, England: Cambridge University Press, 1982.

———. "Confucianism as a Basis for Capitalism." In *Inside the Japanese System: Readings on Contemporary Society and Political Economy,* edited by Daniel I. Okimoto and Thomas P. Rohlen, pp. 36–38. Stanford, Cal.: Stanford University Press, 1988.

Moritani, Masanori. *Japanese Technology: Getting the Best for the Least.* Tokyo: Simul Press, 1982.

Nemoto, Mitsuhiro. "The ERATO Project: Coming Down the Pipeline." *Look Japan* (November 1987): 20.

———. "The ERATO Project: Keep Up the Good Work." *Look Japan* (October 1987): 25.

Nihon Keizai Shimbun, Inc. *The Path Towards a Creative Japan.* Tokyo, May 1988.

"Creativity Is Born from Concentration and Tenacity." *Nikkei Business* (21 July 1986): 128–134.

Nishizawa, Junichi. "A Call for Creative Research." *Japan Echo* 14, no. 2 (Spring 1987). Translation, slightly abridged, of *Nihon no kagaku gijutsu ga abunai,* in *Chuo Koron* (February 1987): 212–216.

———. "Creative Research Activity and the Role of Research Association Journals." *Electronic Data Communications Research Association Journal* 70, no. 10 (October 1987): 967–978.

Nonaka, Ikujiro. "A New Style of Management: Yamaha." *Look Japan* (September 1987): 16–17.

———. "Success the Old-fashioned Way—Company Strategies: Sumitomo Electric." *Look Japan* (October 1987): 16–17.

———. "Toward Middle-Up-Down Management: Accelerating Information Creation." *Sloan Management Review* (Spring 1988): 9–18.

Oka, Akihito. "Internationalization: 1987 White Paper." *Look Japan* (April 1988): 24–25.

Okumura, Akihiko. "Restless Creativity: Kyocera." *Look Japan* (November 1987): 16–17.

Oshima, Tairo. "Brewing the Science of Tomorrow: Protein Engineering." *Look Japan* (November 1988): 3.

Robinson, Brian. "Japan R&D Thriving." *Electronic Engineering Times,* 12 December 1988.

Sakashita, Yuko. "Innovation and Openness: Frontier Research Program." *Look Japan* (July 1987): 22.

Sakurai, Miyoko. "NEC Prepares Princeton Facility." *Electronic Engineering Times,* 4 July 1988.

Sasaki, Hiroshi; Shinya Tsutsui; and Takayuki Yasui. "Japanese Corporate R&D—A Survey of 212 Manufacturing Companies." *Nikkei Business* (4 August, 1 September, and 15 September 1986).

Sasanuma, Osamu. "Human Frontier Science Program: Original and Important." *Look Japan* (September 1987): 15.

"Unprecedented New Research Lab Construction Boom." *Shuukan Diamond* (12 September 1987): 88–95.

Wada, Masami. "Industry and Technology: Fifteen Years Later." *Look Japan* (February 1989): 30–31.

Watanabe, Chihiro. "Foreign Laboratories Move in." *Look Japan* (September 1987): 24–25.

———. "Meeting the R&D Challenge: White Paper on Industrial Technology." *Look Japan* (January 1989): 22–23.

Westney, Eleanor D., and Kiyonori Sakakibara. "The Challenge of Japan-based R&D in Global Technology Strategy." *Technology in Society* 7 (1985): 315–330.
Yoder, Stephen Kreider. "Stifled Scholars: Japan's Scientists Find Pure Research Suffers Under Rigid Lifestyle." *Wall Street Journal,* 31 October 1988.

Chapter 14. Whither Japan in the Twenty-first Century?

Abe, Yoshio. *People-Oriented Management.* Tokyo: Diamond Press, 1987.
"Japanese Enterprises Begin Hiring Foreigners." *Focus Japan* (September 1987): 3.
Hakuhodo Institute of Life and Living. *Japanese Seniors: Pioneers in the Era of Aging Populations.* Tokyo: Hakuhodo Institute of Life and Living, 1987.
Ibuki, Taku. *The Human Resources Revolution: How to Become More Creative.* Tokyo: PHP Publishing, October 1987.
Koinuma, Nobuo. "Challenges of the Graying Society: Japan-U.S. Health Care in Comparison." *Journal of Japanese Trade and Industry,* no. 3 (1988): 43–46.
Misawa, Mitsuru. "New Japanese-Style Management in a Changing Era." *Columbia Journal of World Business* (Winter 1987): 9–17.
Nakatani, Iwao. *The Japanese Firm in Transition.* Tokyo: Asian Productivity Organization, 1988.
"Listening to Top Research Managers." *Nikkei Electronics* (11 January 1988): 73–89.
Ogawa, Akira. "Salaryman Blues." *Look Japan* (July 1988): 7–9.
Omachi, Motohide. "Working Harder at Relaxing." *Journal of Japanese Trade and Industry,* no. 3 (1988): 16–17.
Rohlen, Thomas P. *Japan's High Schools.* Berkeley: University of California Press, 1983.
Science and Technology Agency (STA). *1987 White Paper on Science and Technology.* Tokyo: Science and Technology Agency, 1987.
Takemura, Kenichi. *The Era of Genius.* Tokyo: Gakushu Kenkyusha, 1987.
Takenaka, Heizo. "Fewer Jobs, Less Loyalty." *Look Japan* (May 1988): 20.
Takesue, Takahiro. "Foreigners Finding a Niche in Japanese Firms." *Journal of Japanese Trade and Industry,* no. 2 (1988): 25–27.
Uchimura, Takashi. "A Job for the Headhunters." *Journal of Japanese Trade and Industry,* no. 6 (1987): 27–29.
Walters, Donna K. H. "Job-jumping Loses Its Stigma for Japanese." *Los Angeles Times,* 9 July 1987.
White, Merry. *The Japanese Educational Challenge.* New York: Free Press, 1987.

INDEX

Abacus, 58
Adaptive creativity, 16–17, 49
Advanced-compatible television (ACTV), 143
Advanced Micro Devices (AMD), 24, 156
Advanced television (ATV), 144; *see also* High-definition television
Advice on Nonduality (Mori), 42
Aesthetics, 55–59
After-hours socializing, 89–90, 93–94
Agency for Industrial Science and Technology (AIST), 233, 238
Aging population, 263
Akihabara, 107–8, 149–50
Akutsu, Takashi, 202
Alpha 7000 camera, 28–29
Alps Electric, 250
American Electronics Association (AEA), 4, 144
American National Standards Institute (ANSI), 144
Ampex, 117
ANNIE (Applications of Neural Networks for Industry in Europe), 188
Aoki, Sadami, 109
Apple Computer, 249
Arback Seihaku, 35
Architecture, 57–58, 91–92
Artificial intelligence (AI), 59, 66, 75, 76, 168, 171, 173–74, 175–76, 190–91, 268, 269
ASCII Corporation, 23
ATR Interpreting Telephone Project, 180–81, 239–40
Audio technology, 82–84
Automatic Data Processing System (ADPS), 178
Automax, 43
Automobiles, 15, 38, 268; designers, 58; design-your-own, 22–23, 125; electric, 211

Basic research, 4, 6, 7, 9, 46–47, 53, 54, 171–72, 221–24, 235–36
Batteries, see-through, 11
Bayer, 3
Bednorz, J. Georg, 194
Bell Labs, 184, 195
Bioceramics, 56, 67
Biochemistry, 9
Biochips, 192, 193
Biocommunications, 67
Biocomputers, 66, 190–92, 238
Bioelectronics, 9, 66, 191–92, 237–38
Bioengineering, 56
Bioholonics, 190
Bio-homeostatis, 237–38
Biosensors, 8, 56, 66, 191, 193
Biotechnology, 11, 66, 240, 271–72
Bonsai, 56, 81
Bosch, Robert, 143
Brainstorming, 104, 106, 109–13
Breadmaker, automatic, 11, 68–69, 255, 262
Breakthrough(s): creativity, 16–17, 49; generating, 103–15
British Broadcasting Corporation (BBC), 143
Buddha in the Robot, The (Mori), 42
Buddhism, 21, 41–48, 52
Buyouts, 256, 257

Calculators, 78, 120, 151–52, 153
California Institute of Technology, 259

Cameras, 27–29, 121–22, 123
Campbell, Gordon, 23
Canon, 3, 24, 29, 63
Carter, Forrest, 191
Cartesian logic, 21–22
Casio, 78, 178
Cava, Robert J., 81
CD-ROM disks, 82
Cellular telephones, 150
Ceramics, 61, 95, 164
Charge-coupled devices (CCDs), 28, 68, 133, 184
Chemistry, 66
China, 10, 24, 48, 58, 147
Chips, 4, 23–24, 66, 71, 90, 108, 144, 156, 158, 182, 246–47, 269; "sets," 23; superconducting, 211–13, 241–42
Chips and Technologies, 23–24
Chu, Ching-Wu "Paul," 81, 195–96
Clear Vision television, 142
Color graphic terminals, 130
Columbia University, 259
Communist bloc, 10
Compact disk (CD) players, 6, 10, 68, 82–83
Computer-aided design (CAD), 56, 59, 66, 107, 124, 170, 212, 269
Computer-integrated manufacturing (CIM), 21
Computers, 63, 66, 75–76, 90, 118, 130, 238; advanced image, 138; future, 149–93, 270; and HDTV, 136, 138, 145–48; superconducting, 211–13, 241–42; translation systems, 59
Confucianism, 47–48
Conglomerates, 93
Consensus building, 33
Convex, 31
Copiers, 63, 92, 120–21, 268
Copy-Jack, 120
Copyrights, 10, 108, 272
Corporate: research, 242–60; restructuring, 30–31
Cosmo AT, 35
Council for Science and Technology, 234
Cray Research, 161, 165, 167
Creative fission, 19–21
Creative fusion, 19–21, 23–25
Creativity: circles, 106–7; cultivated, 17–19; encouraging, within companies, 33–34; -inducing environments, 90–92; mandala of, 50–55; training, 104; *see also* Japanese creativity
Crossover technologies, 67–68

Daewoo, 10
Daiwa House, 124
Dambrot, Stuart M., 193
Danby, J. R., 205
Data-processing technology, 59
Dataquest, 4, 24
David Sarnoff Research Center, 143
Da Vinci, Leonardo, 51
Debevoise, Malcolm, 232
Decline and Prosperity (Makino), 88
Defense Science Board, 3
Del Rey Group, 143
Denicoff, Marvin, 171–72, 175
Dentsu, 147
Design arts, 46, 58
Design-your-own-home systems, 124
Devol, George C., Jr., 17
Digital: audio technology, 11, 82; chips, 133; circuits, 133; sound-field processor, 107; sound technology, 147
Direct broadcast satellite (DBS), 134–35
Diversification, 31, 34
Dot matrix printers, 79
DRAM (dynamic random access memory), 90, 257–58

Eastman Kodak, 260
Education, 6–7, 160–61, 222, 223, 224–32, 264–65
8mm video cameras, 6, 68, 118
Electrical power, 175, 211; storage systems, 213–15

Electronic: copyboard, 63, 92; dictionaries, 155; publishing, 138
Electronics Industries Association of Japan (EIAJ), 154
Electronics industry, 9, 149, 158, 240, 243; evolution of Japan's, 6, 56, 68; ideas, 114; and superconducters, 211–13; U.S., 143–44
Elkus, Richard J., Jr., 145
Employee: flexibility, 80; and ideas, 122–23; small business, 35; stress, 44–45; training, 18
Energy Conversion Devices (ECD), 78
Energy systems, 214–15
Engelberger, Joseph, 17
Engineering Research Association of Optoelectronics Applied Systems, 182
Era of Genius, The (Takemura), 113
Esaki, Leo, 220, 232
Essence, extracting, 68–69
EUREKA 95, 143
European Strategic Programme for Information Technology (ESPRIT), 188
Expert systems, 176
Exploration/search (*tansaku*), 50, 53, 54, 73–85
Exploratory Research of Advanced Technology (ERATO), 233, 236, 237
Extended-definition television (EDTV), 137, 141, 142

Facsimile machines, 6, 11, 15, 59, 63
Factory automation software, 16
Fairchild, 32
Federal Communications Commission (FCC), 143–44
Feed-forward method, 113–14
Feigenbaum, Edward A., 171
Fernbach, Disney, 165
Fifth Generation, The (Feigenbaum and McCorduck), 171
Fifth-generation computers, 136, 150, 155, 161, 164, 167–76, 178–79, 182, 192, 203, 238
Fifth Generation Fallacy, The (Unger), 76, 168
Finance-service fusion, 37
Fine Ceramics Product R&D Institute, 61
Folding fans, 58
Folk arts, 55–59, 119
Foreign: companies, 202–4; condescending attitude toward, 266; R&D centers, 244–47; R&D in Japan, 247–49; researchers, 229–32, 233–34, 254; universities, 259–60
Four-dimensional workstations, 124
France: Ministry of Public Posts and Telecommunications (PPT), 153; SECAM system, 131
Frontier: materials, 237, 238, 240; Research program, 236–38
Fuchi, Kazuhiro, 173, 238
Fueki, Kazuo, 195
Fuji Photo Film, 121–22, 260
Fujitsu, 20, 21, 24, 32, 56, 75, 108, 133, 155, 156, 159, 161, 162, 165, 171, 176, 177, 181, 184, 188, 192, 197, 212, 236, 246, 252–53, 254, 257
Fukui, Kenichi, 225
Funai Electric, 11, 68
Fusion, creative, 19–21; of East and West, 22–24; global, 24–25
Fusion (*yugo-ka*), 20; Law (1988), 66; technologies, and recycling, 65–67; "technology," and small business, 37–38
Fusion Systems Corporation, 271, 272
Fuzzy Engineering Research Laboratory, 190–91
Fuzzy logic, 21–22; computer, 9, 187, 189–90, 269

GaAs (Gallium arsenide), 162, 236
Galbraith, John Kenneth, 273
General Accounting Office (GAO), 4
General Electric, 3
Generation of ideas, 50, 53, 54
Germany: PAL system, 131
Glenn, William E., 143

Global: economy, 14–15, 113, 167; environment, 247; fusion, 24–25; search, 76–78
Goguen, Joseph, 172
Gordon, W. J. J., 104
GRAPHICA, 35, 254
Group creativity, 15, 19, 48, 52–53, 109
Guarded Horn Clause (GHC), 175
Hotels, capsule, 69
Houston Research Council, 4
Human Frontier Science Project, 234
Humanware industry, 270
Hybrid: "innovation" services, 36; technologies, 63–64
Hypermedia, 138, 147, 150
Hyundai, 10

Haiku, 121
Hamamatsu Photonics, 187
Hanae Mori, 35
Handicapped, 158–59
HD-MAC system, 145
Head-hunting, 251–54
Hewitt, Carl, 172
High-Definition System for North America (HDS-NA), 143
High-definition television (HDTV), 4, 16, 63, 69–70, 114, 129–48, 247, 258, 268
High-electron mobility transistor (HEMT), 162–63
"High-rise" chips, 71
High-sense, 263
High-speed signal processors, 133
Hinduism, 52
Hirai, Katsuhide, 20, 21
Hiraiwa, Gaishi, 202, 247
Hiring, 249–51
Hirohito, Emperor, 270
Hitachi, 3, 4, 10, 11, 24, 31, 56, 75, 108, 133, 142, 155, 156, 159, 161, 162, 165, 170, 171, 173, 175, 177, 181, 192, 197, 208, 212, 228, 236, 246, 258
Hi-Vision television, 126–43, 144; Promotion Council, 140
Holographic video-conferencing system, 151
Home electronics, 68
Homma, Keiichi, 227
Honda, Soichiro, 42, 43, 250
Honda Motors, 10, 42, 43, 250–51
Hopfield, John J., 185
Hotate, Kazuo, 183

IBM, 23, 80, 156, 173, 181, 194, 212, 248–49, 252, 253
Ibuka, Masaru, 70
Idea: contests, 43; evolution of Japanese, 54; mandala of, 50, 52–53, 54; *see also* Brainstorming
Ido, Yoshimitsu, 32, 33
Ihori, Shigeo, 113
Imai, Kenichi, 97
Imaizumi, Yoshihisa, 92
Improved-definition television (IDTV), 141, 142
Inamori, Kazuo, 32
Incompleteness, 122–23
Individual creativity, 15, 16, 17, 19, 33, 47, 48, 52–53, 264, 268
Industrial robots, 17
Industry: heavy, 30–31; "infant," 36
Information Network System (INS), 178
Information Processing Association (IPA), 177–78
Initial public offering (IPO), 36
Innovation, 53
Institute of Industrial Technology, 187
Institute of Next-Generation Computer Technology (ICOT), 169, 170–76, 238–39
Integrated Services Digital Network (ISDN), 179
Intel, 108, 156, 258
International Leisure Expo, 141
International Solid State Circuits Conference, 3, 90
International Superconductivity Technology Center (ISTEC), 202–4, 234

Intuition, 45–46
Ishida, Tokuji, 28
Ishikawajimi-Harima Heavy Industries (IHI), 31
Isuzu Motors, 250

Japan as Number One (Vogel), 76
Japan Broadcasting Corporation (NHK), 110, 129–35, 140–41, 142, 143, 144, 145, 146
Japan Economic Journal, 35, 93, 122, 175, 199, 221, 243
Japan Electronic Dictionary Research Center, 155
Japanese companies: and creativity, 33–38, 52–53, 242–43; and overseas companies, 31–32; and R&D, 242–60; small, 34–38; women in, 265–66
Japanese creativity: barriers to, 264–66; and changes, 11–12; within companies, 33–38, 52–53, 242–43; danger of denying, 7–9; drive for, 29; in electronics industry, 6; forces behind, 10–13, 37; and future, 261; and hiring, 249–51; origins of, 41–60; rise of, 4–7; in science, 219–24; and smart cards, 154; vs. Western creativity, 15–23
Japanese gardens, 56–57, 122
Japanese government: and HDTV, 136–41; protection of infant industries, 18; and research, 75, 232–40; and small business, 36; and supercomputers, 166; and superconductors, 199–204
Japanese language, 8–9, 59, 76, 79, 123, 170, 175–76, 230–31
Japanese Technology (Moritani), 222
Japanese Technology Evaluation Program (JTECH), 171–72
Japan Industrial Journal, 223
Japan Railway, 204–8
Japan Research and Development Corporation (JRDC), 190, 236
"Japan's American Genius" (Ovshinsky), 78
Japan Society for the Promotion of Science (JSPS), 233
Japan Television and Broadcasting Association (JTBA), 142
Jobs, Steven, 146
Josephson, Brian, 163
Josephson junction devices, 163–64, 212
Journal of Japanese Trade and Industry, 92

Kakidani, Hitoshi, 227
Kamiya, Akimi, 255
Kanahara, Kazuo, 156
Kanji writing system, 11, 59, 124, 159, 168; *kana*-conversion, 79
Kansai Cultural Research City, 57
Kao Corporation, 11, 33–34
Kapor, Mitch, 15, 111
Karuta card game, 105*n*
Kawakami, Hiroshi, 71
Kawakami, Tetsuro, 202
Kawakita, Jiro, 104, 105
Kawasaki Steel, 30
Key Technology Center (KTC), 240–42
Kibi Highlands Technopolis, 57
Kick in the Seat of the Pants, A (von Oech), 50
Kikuchi, Makoto, 16, 73
Kikusui Homes, 124
Kishimoto, Yoriko, 35, 93
Kitahara, Yasusada, 105–6
KJ (Kawakita Jiro) method, 104–6
Knowledge base, 168–69, 176
Kobe Steel, 30
Kodama, Fumio, 65, 67
Koot, Cees, 145
Koyama, Hiroyuki, 251
Kreisman, Norman H., 166
Kuhn, Thomas, 48
Kurata, Masahi, 92
Kurata, Ryoichi, 23
Kyocera, 23, 32, 35, 61
Kyotani, Yoshihiro, 204, 205, 206, 207–8
Kyoto, 94; Center for Japanese Studies (KCJS), 231

Language barrier, 8–9; *see also* Japanese language
Laptop computers, 6, 58, 79, 118
Lee, Chris, 226, 227
Lee, O-Young, 56, 119, 121
Light-emitting diode (LED), 120, 184
Linear motor cars, 204–8
Linear reasoning, 21
Liquid crystal displayers (LCDs), 79, 118, 149
List Processing (LISP), 169, 176
Local area networks (LANs), 155
Logical inference machine (LIM), 175
Look Japan, 220
Lotus blossom technique, 110–11, 154
Lotus 1-2-3, 15, 21, 111
LSI Logic K.K., 18

McCorduck, Pamela, 171
McCulloch, Warren S., 185
Made in Japan (Morita), 8, 70
Maglev technology, 204–6
Magnetic Resonance Imaging (MRI), 211
Makino, Noboru, 88
Management, 88, 105, 242; and technology fusion, 66–67
Managing Automation, 193
Mandala: approach to creativity, 50–55, 268; of HDTV, 148
Manufacturing, 5–6, 21
Marketing fusion, 37
Market-share battles, 107–8, 216
Maruta, Yoshio, 34
Massachusetts Institute of Technology (MIT), 259
Masscomp, 31
Matrix frameworks, 80–81
Matsubara, Sueo, 43, 47, 53
Matsuo, Takayuki, 37
Matsushita, 10, 11, 63, 68–69, 70, 75, 117, 120, 133, 142, 155, 156, 158, 197, 247, 255, 262
Maxxum 7000, 28–29, 123
Mazda, 4, 22, 57–58, 125
Media, Japanese, 15
Medical: expert systems, 75, 170, 268; field, and superconductors, 211
Meditation, 45, 46
Memory chips, 24, 32, 56, 69, 90, 133, 136, 242, 268; video, 142
M-Form Society, The (Ouchi), 89
Microprocessors, 24, 108, 155, 258, 268
Miniaturization, 17, 55–56, 119–21, 152
Ministry of Education (MOE), 199, 201–2, 212–13, 225, 226, 228–29, 230, 231, 235, 272
Ministry of International Trade and Industry (MITI), 35, 36, 71, 89, 96, 98, 99–101, 158, 229, 233–36, 238–39, 240, 272; Computer Education Center, 160; and computers, 182, 183, 184, 187–88, 190, 191–92; Fifth Generation Computer Project, 75–76, 155, 167, 172, 203; and HDTV, 136, 137–38, 140; Industrial Structure Deliberation Council, 65–66, 75; Small Business Fusion Promotion, 37–38; Supercomputer Project, 108, 161–65, 200, 211–12; and superconductivity, 195, 196–98, 199, 200, 202–3, 211–12, 215; Venture Enterprise Center (VEC), 94–95; vision statements, 74–75
Ministry of Posts and Telecommunications (MPT), 135, 136–38, 140, 144, 145, 179–80, 199–200, 202, 240
Ministry of Transport (MOT), 202, 208
Minolta, 4, 26–28, 123
Minsky, Marvin, 185
MIPS Computer, 24
Misawa, Mitsuru, 33
Mita, Katsuhige, 202
Mitsubishi, 58, 75, 88, 109–10, 155, 156, 159, 161, 162, 170, 171, 173, 175, 177, 187, 192, 197, 208, 258, 271, 272
Mitsui and Company, 23
Mizukami, Mitsuo, 66
Mizuno, Denichi, 190
Mori, Kiyoji, 214

Mori, Masahiro, 42, 44
Morishima, Michio, 47
Morita, Akio, 8, 70
Moritani, Masanori, 222
Motorola, 24, 156, 258
Mueller, K. Alex, 194
Mukaibo, Takashi, 215
Mukta Institute, 41–44, 47, 53
Multicore Project, 201
Multidisciplinary product development teams, 68
Multi-functional creativity, 22–23
Multimedia, 150, 170
Multinational companies, 247–49
Multiple: -activity thinking, 113; -track development, 107–8; use, 17
MUSE (multiple sub-Nyquist sampling encoding) system, 134, 140, 143, 145
Muto, Yoshio, 202
MY (Matsumura Yasuo) Method, *see* Lotus blossom technique

Naito, Masami, 192
Nakagin Capsule Tower Building, 125
Nakajima, Sadao, 202
Nakamura, Hajime, 46
Nakasone, Yasuhiro, 266
Nakayama, Keiji, 121
National, 156
National Academy of Engineering, 185
National Information System for Science and Technology (NIST), 238
National Rehabilitation Center, 158
National Science Foundation (NSF), 4
National Space Development Agency (NASDA), 134
National Television System Committee (NTSC), 131, 134, 137, 141, 142
Naya, Mikio, 26–27, 28, 38
Nebashi, Masato, 89–90
NEC, 4, 18, 24, 56, 108, 114, 121, 124, 133, 155, 156, 160–61, 162, 165, 170, 171, 177, 181, 188, 212, 236, 247, 253, 254, 255–56, 258
Neural networks, 9, 185–87, 191
Neurocomputers, 185–89, 191, 193
New Business Association, 36, 37
"New Media Community Cities," 138
New products, 10, 35, 120–21; from Fifth Generation, 172–73; of 1990s, 268–70; and old, 62–69; superconductors and, 208–11
New ventures, profitability of, 34
New York Institute of Technology (NYIT), 143
NeXT computer, 146
Nikkei AI, 175
Nikkei Business survey, 34
Nikkei Electronics, 139, 268
Nikon, 29, 123
Nintendo, 250
Nippon Credit Bank, 34
Nippon Kokan (NKK), 30, 31–33
Nippon Steel, 30, 31
Nishimura, Yoshifumi, 223
Nishizawa, Junichi, 221, 236
Nissan, 10, 24, 252
Nonaka, Ikujiro, 34, 251
Nondualism, 52
NTT Corporation, 150, 151, 154, 155, 156, 167, 178–79, 181, 197, 211
Numano, Yoshinobu, 141
Nurturing (*ikusei*), 18, 50, 53, 54, 87–101

Oeki, Masanori, 205
Office environment, 91–92
Office of Technology Assessment (OTA), 198, 216
Okamoto, Michio, 228
Oki Electric, 56, 63, 155, 156, 162, 170, 177, 181, 257
Olympus, 29
Onishi, Minoru, 122
Optical character recognition (OCR) system, 188
Optical Technology R&D Corporation, 183–84

Optocomputers, 9, 59, 66, 181–85, 193
Optoelectronic integrated circuits (OEICs), 183, 184
Optoelectronics, 4, 136, 182–83
Optomechatronics, 67
Osborne, A. F., 104
Osuga, Katsumi, 144
Ouchi, William, 89
Ovshinsky, Stanley, 78

Paper folding (*origami*), 58
Papert, Seymour, 185
Paradigms, 48–49
Parallel processing, 75, 162, 164, 168, 171, 173–74, 175, 187
Particle accelerators, 211
Patents, 3, 4, 10–11, 108, 197–98, 202–3, 234, 258, 271–72
Pentax, 29, 123
Personal computers, 118
Personal sequential inference (PSI), 175
Philips NV, 143, 145
Philosopher-managers, 34
PIMOS (parallel inference machine operating system), 175
Pitts, Walter, 185
Plus and Company, 63, 92, 120
Pocket computers, 149–50, 151–54, 155
Poincaré, Jules-Henri, 15
Postdoctoral programs, 226–27
Powell, G., 205
Process innovation, 16, 17; spin-offs, 69–70
Product: design, 5; development, 33–34, 79; innovation, 7, 16; *see also* New products
PROLOG (Programming in Logic), 169, 170, 175
Protein Engineering Research Institute (PERI), 191, 238
Pulse-code modulation (PCM), 82, 135
Purity, 52

Quality, 5, 6, 21; circles (QC), 20, 106–7

Rational thought, 45, 47, 49
RCA, 131, 184
Recycling (*sairiyo*), 50, 54, 55, 61–71
Reduced instruction set controller (RISC) systems, 24
Refinement (*kaizen*), 50, 53, 54, 80, 117–26
Regional research core program, 98–101, 229
Relational data bases, 75, 169, 171
Research and development (R&D), 94–95, 97, 242–60; creative, 20, 73, 75, 76; spending, 11, 235, 239
Researchers, 218–60
Reverse thinking, 113
Richardson, John, 166
Ricoh, 63, 91
Ridge Computer, 31
Riggs, Henry E., 16
Right- vs. left-brain thinking, 46
Ringi system, 33
Robots, 17, 37, 42, 43, 68, 136, 188, 193, 268, 269
Rochester University, 260
Rosenblatt, Frank, 185

Sakamura, Ken, 155, 159
Samsung, 10
Sandwiching ideas, 70–71
San Jose Mercury News, 145
Sanyo Electric, 11, 66, 91
Sasaki, Tadashi, 78
Satellite broadcasting, 133, 134–38, 140, 141, 142, 145; *see also* Direct broadcast satellite
Sato, Sosuke, 271
Satori, 45, 52, 53
Scanning devices, 59, 121
Science: in Japan, 219–24; Western, and Japanese philosophy, 46–48, 53–54
Science and Technology Agency (STA), 195, 199, 200–1, 212, 219,

220, 225, 233, 234, 235–36, 238, 269–70
Scientific Computer Research Association, 162
Seiko, 56
Semiconductor Manufacturing Technology (SEMATECH) consortium, 144
Semiconductors, 4, 23–24, 32, 38, 130, 246, 257–60; and future computers, 156, 241; and HDTV, 133, 144; lasers, 184
Seventh-generation computing, 192–93
Sharp, 78, 81–82, 83, 120, 152, 155, 181, 192, 197
Shikano, Kiyoshiro, 180
Shima, Keiji, 130
Shimizu, Hiroshi, 191
Shintoism, 47
Siemens, 24
Sigma project, 176–78
Silicon Valley, 4, 8, 23, 95, 146
Silk Road Expo, 141
Silver Phones, 150
Simplicity, 121–22
Singapore, 10, 48
Sloan Management Review, 251
Small businesses, 34–38
Smaller Is Better (Lee), 56, 119
Smart cards, 151, 153–54
Social transformations, 262–64
Socratic method, 45
Sodec, 35
Software, 10, 21, 32, 58, 66, 76, 139–40, 154, 176–78, 188, 250; automatic creation, 170, 177–78; multilingual, 158, 170; for supercomputers, 162, 166; video, 147
Solar: calculator, 78, 120; cells, 11
SONLI, 176
Sony, 4, 8, 10, 24, 32, 35, 56, 63, 68, 70–71, 73, 114, 117, 118, 123, 133, 142, 144, 146, 150, 181, 192, 246–47
SOR (synchrotron orbital radiation), 212–13, 242
Sord Computer, 35
Sortech, 214, 241
Sotoyama, Kazumi, 207
South Korea, 10, 48, 63, 138
Soviet Union, 11, 48, 79
SPARC microprocessor chip, 24, 246
Spin-off, 130, 239; process, 69–70
Spinouts, 256
Spiral development, 78–80
Spontaneous creativity, 17–19
SQUIDs (superconducting quantum interference devices for medical sensors), 210, 212
Steel industry, 30–31
Stereo televisions, 63
Stock market crash, 62
Strategic alliances, 256–59
Stress, 44–45
Structure of Scientific Revolutions, The (Kuhn), 48
Studioman, 150
Suijaku mandara, 52
Sumita, Shoji, 206
Sumitomo Industries, 30, 197–98, 211, 214, 236
Sun Microsystems, 24, 246
Supercomputers, 4, 108, 136, 161–67, 182, 211, 268
Superconductivity, 80–81, 108, 118, 164, 194–217, 229, 268, 269
Superconductivity Generator Materials and Research Association (Super-GM), 214
Superminicomputers, 31, 32
Suzuki, Daisetz, 45
Suzuki, Naomichi, 240
Synectics, 104
Synergistic research, 37
Systems research, 16

Taishi, Shotoku, 34
Taiwan, 10, 48, 138
Takagaki, Takashi, 253
Takahashi, Hiroshi, 104
Takei, Fumihiko, 202
Takemura, Kenichi, 113
Takeshita, Noboru, 24

Tanahashi, Yuji, 187
Tanaka, Shoji, 194, 199, 200, 202, 229
Tashima, Hideo, 27
Tateishi, Kazuma, 42, 43
Taxin, Harry, 146
Teamwork, 16, 19, 73, 106
Techno-globalism, 48, 270–73
Technology: fusion, 37, 55, 65–67, 103; -marketing fusion, 37; road maps, 82–84, 105–6; -service fusion, 37; trees, 81–82, 83
Technomart programs, 97
Techno-nationalism, 48, 55
Technopolis program, 95–98
Technovalley Intelligent Core, 98
Telecommunications, 10, 185
Telematique Plan, 153
Telephone, 15; cards, 154; translation, 179–80, 188, 239
Television, 90, 129–48; broadcasting, 137, 142; and computers, 146–47; screens, 133; spin-offs, 70
TEMAKI speakers, 58
Textiles, 63
Thermal printers, 79
Thin film, 211
Thinking: in circles, vs. straight, 22, 106–7; Confucian, 47–48; original, 91; rational vs. Zen, 45–46
32-bit microprocessors, 156, 158, 258
Time magazine, 5
Togai, Masaki, 189
Togano, Kazuma, 196
Tohoku Semiconductor, 24
Tonegawa, Susumu, 218–19, 220, 224, 225
Toshiba, 3, 4, 11, 24, 63, 79, 114, 118, 121, 135, 142, 155, 156, 162, 170, 175, 177, 181, 184, 197, 208, 258–59
Totani, Tatsuo, 252
Toyota, 10, 22, 43
Transformation, 125–26
Transformer toys, 22, 125
Translation, 175, 176, 188, 238, 269; multilingual system, 170; phones, 9, 150, 179–81, 239; voice, 179
TRISTAN (Transposable Ring-Intersecting Storage Accelerator in Nippon) ring, 210–11
TRON (The Real Operating Nucleus) Project, 155–61, 258, 268
Tsugiura, Hideo, 250
Tsujikawa, Ikuji, 202
Tsukuba Science City, 57, 95, 184, 232–33; University, 228–29
21st Century Plaza, 98

Uchida, Motokazu, 133
Uchida, Shinichi, 196
Ueda, Kazunori, 175
Uenohara, Michiyuki, 247, 255–56
Ultrafast chip technologies, 108
Unger, J. Marshall, 76, 168
Uni-functional creativity, 22–23
United States: broadcasting industry, 131; computer technology, 188; feedback mentality, 114; and HDTV, 143–46; ignorance of Japanese technology, 8–9; industrial policy, 18–19; and Japanese companies, 23–24, 31–32; vs. Japanese search approaches, 77; and optical technology, 184–85; research, 235; supercomputers, 166–67; and superconductors, 197–99, 216–17; technological lead, 6–7, 261; universities, 231
U.S. Department of Defense, 144
U.S. Department of Energy, 4
U.S. Navy, 3
U.S. Patent and Trademark Office, 3, 272
U.S. Super 301 trade law, 272
Universities: foreign, 259–60; research, 224–32
User-friendly terminals, 169–70

Venture capital, 35–37, 93–101
Venture Enterprise Center, 36

Video: bandwidth compression, 133, 134; cameras, 132; cassettes, 10; computing, 132, 146–47, 150, 268; discman, 63; disks, 130; games, 21; phones, 63, 150, 151, 180; RAM chips, 133; tape recorders (VTRs), 63, 70, 90, 117, 129, 130, 132, 147
Videocassette recorders (VCRs), 114, 138
Visionary thinking, 74–76
Visual thinking, 106, 123–24
Vitelic, 32
VLSI (very large-scale integrated) technology, 24, 75, 89–90, 107
Vogel, Ezra, 76
Voice: -activated word processors, 79, 181; recognition, 188; translation, 179
Von Neumann, John, 168
Von Oech, Roger, 50

Walkman, 63, 70–71, 118
Wall Street Journal, 5
Watanabe, Hiroyuki, 189
Watanabe, Itaru, 218
Western creativity, 15–23, 48–50, 53–55
Western culture, 32
West Germany, 188, 235
Why Has Japan Succeeded? (Morishima), 47
Women, 11, 68, 262–63, 265–66, 267
Wood carving (*netsuke*), 55, 57
Word processors, 6, 59, 79, 181
Wordstar, 21
Wysocki, Bernard, Jr., 5

Xerox, 3

Yagi Antenna, 135
Yamaha, 11, 107
Yamakawa, Retsusuke, 189
Yasukochi, Ko, 202
Yasuo, Matsumura, 111
Yawata, Keisuke, 18
Yoder, Stephen Kreider, 223
Younger generation (*shinjinrui*), 11–12, 91, 223, 254, 262, 267

Zadeh, Lotfi, 189
Zen, 34, 42, 43, 44, 45–48, 58, 121, 122

ABOUT THE AUTHOR

SHERIDAN M. TATSUNO is the principal of NeoConcepts, a high-tech consulting firm in Fremont, California. Prior to forming his own company, he had thirteen years of experience in market research, planning, and international finance with Dataquest, Bechtel Corporation, and Woodward-Clyde Consultants.

Mr. Tatsuno received a B.A. degree in political science from Yale University in 1972 and an M.A. degree in planning and policy analysis from Harvard University's Kennedy School of Government in 1977. In addition to these credentials, Mr. Tatsuno speaks Japanese, French, and Spanish and has authored a previous book, *The Technopolis Strategy: Japan, High Technology, and the Control of the Twenty-first Century.*